Selected Titles in This Series

(Continued in the back of this publication)

MEMOIRS
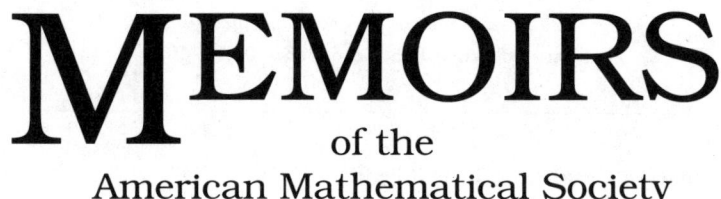

of the
American Mathematical Society

Number 605

Classification of Simple C^*-algebras: Inductive Limits of Matrix Algebras over Trees

Liangqing Li

May 1997 • Volume 127 • Number 605 (first of 4 numbers) • ISSN 0065-9266

American Mathematical Society
Providence, Rhode Island

1991 *Mathematics Subject Classification.*
Primary 46L05, 46L35; Secondary 19K14, 46L80.

Library of Congress Cataloging-in-Publication Data

Li, Liangqing, 1959–
   Classification of simple C^*-algebras : inductive limits of matrix algebras over trees / Liangqing Li.
      p. cm. — (Memoirs of the American Mathematical Society, ISSN 0065-9266 ; no. 605)
   "May 1997, volume 127, number 605 (first of 4 numbers)."
   Includes bibliographical references
   ISBN 0-8218-0596-7 (alk. paper)
   1. C^*-algebras. I. Title. II. Series.
QA3.A57   no. 605
[QA326]
510 s—dc21
[512′.55]

97-421
CIP

Memoirs of the American Mathematical Society

This journal is devoted entirely to research in pure and applied mathematics.

Subscription information. The 1997 subscription begins with number 595 and consists of six mailings, each containing one or more numbers. Subscription prices for 1997 are $414 list, $331 institutional member. A late charge of 10% of the subscription price will be imposed on orders received from nonmembers after January 1 of the subscription year. Subscribers outside the United States and India must pay a postage surcharge of $30; subscribers in India must pay a postage surcharge of $43. Expedited delivery to destinations in North America $35; elsewhere $110. Each number may be ordered separately; *please specify number* when ordering an individual number. For prices and titles of recently released numbers, see the New Publications sections of the *Notices of the American Mathematical Society.*

Back number information. For back issues see the *AMS Catalog of Publications.*

Subscriptions and orders should be addressed to the American Mathematical Society, P. O. Box 5904, Boston, MA 02206-5904. *All orders must be accompanied by payment.* Other correspondence should be addressed to Box 6248, Providence, RI 02940-6248.

Copying and reprinting. Individual readers of this publication, and nonprofit libraries acting for them, are permitted to make fair use of the material, such as to copy a chapter for use in teaching or research. Permission is granted to quote brief passages from this publication in reviews, provided the customary acknowledgment of the source is given.

Republication, systematic copying, or multiple reproduction of any material in this publication (including abstracts) is permitted only under license from the American Mathematical Society. Requests for such permission should be addressed to the Assistant to the Publisher, American Mathematical Society, P. O. Box 6248, Providence, Rhode Island 02940-6248. Requests can also be made by e-mail to `reprint-permission@ams.org`.

Memoirs of the American Mathematical Society is published bimonthly (each volume consisting usually of more than one number) by the American Mathematical Society at 201 Charles Street, Providence, RI 02904-2294. Periodicals postage paid at Providence, RI. Postmaster: Send address changes to Memoirs, American Mathematical Society, P. O. Box 6248, Providence, RI 02940-6248.

Contents

ABSTRACT

In this paper, it is shown that the simple unital C^*-algebras arising as inductive limits of sequences of finite direct sums of matrix algebras over $C(X_i)$, where X_i are arbitrary variable trees, are classified by K-theoretical and tracial data. This result generalizes the result of George Elliott of the case of $X_i = [0,1]$. The added generality is useful in the classification of more general inductive limit C^*-algebras.

Key words and phrases: K-theory, traces, classification, simple C^*-algebras, inductive limits. AMS subject classification number: Primary: 46L05.

Received by the editor July 12, 1995.

vi

To my family

Chapter 1

Introduction

In the 1970's, Ola Bratteli [Br] intensively studied AF algebras — inductive limits of finite dimensional C^*-algebras, and obtained a classification theory based on the Bratteli diagram (see also the earlier work [Gl] and [Di]). Shortly afterwards, George Elliott completely classified AF algebras by scaled ordered K-theory.

Motivated by the classification of AF algebras, Effros questioned whether suitable invariants could be found for an inductive limit

$$A_1 \xrightarrow{\phi_{1,2}} A_2 \xrightarrow{\phi_{2,3}} A_3 \xrightarrow{\phi_{3,4}} A_4 \longrightarrow \cdots$$

with $A_n = \oplus_{i=1}^{k_n} M_{[n,i]}(C(X_{n,i}))$, where the spaces $X_{n,i}$ are finite CW complexes and $[n,i]$ are positive integers.

The study of this class of C^*-algebras becomes an essential ingredient of George Elliott's classification program whose ambitious goal is a classification theory of separable nuclear C^*-algebras.

It has been proved by Bratteli, Elliott, Evans, Kishimoto and Putnam that

1

many C^*-algebras, including irrational rotation C^*-algebras can be expressed as inductive limit algebras of the above form, even though it is not at all obvious that they can be expressed as such inductive limits (see [BrEK] [BrK] [EE] [P]).

In 1991, Elliott classified the above class of C^*-algebras for the case of $X_{n,i} = [0,1]$, by invariants consisting of scaled ordered K_0 group, tracial state space TA and the pairing of K_0A and TA (see [El1]).

In this paper we will extend the above result to the case of $X_{n,i} = $ trees. Our main theorem is the following.

Theorem *Suppose that simple unital C^*-algebras A and B can be written as unital inductive limits of finite direct sums of matrix algebras over $C(X_i)$, where the spaces X_i are trees. Suppose that there is an isomorphism of ordered groups*

$$\phi_0 : \ K_0A \longrightarrow K_0B$$

taking $[\mathbf{1}] \in K_0A$ into $[\mathbf{1}] \in K_0B$, and that there is an isomorphism between compact convex sets

$$\phi_T : \ TB \longrightarrow TA,$$

where TA and TB denote the simplices of tracial states of A and B respectively. Suppose that ϕ_0 and ϕ_T are compatible, in the sense that

$$< \tau, \phi_0 g > = < \phi_T \tau, g >, \qquad\qquad g \in K_0A, \ \tau \in TB.$$

It follows that there exists an isomorphism $\phi : \ A \longrightarrow B$ giving rise to ϕ_0 and ϕ_T.

It should be pointed out that, the assumption that ϕ_0 and ϕ_T are compatible is not redundant. We will present two C^*-algebras in the class which have isomorphic scaled ordered K groups and spaces of tracial states, but there is no pair of isomorphisms between the K-groups and the tracial state spaces which are compatible. Therefore, the C^*-algebras are not isomorphic.

The main difficulty arising is that trees do not have total order like the order of interval $[0,1]$. The proof of the case of $[0,1]$, both the existence and the uniqueness, heavily depends on the total order of $[0,1]$. Essentially, we use Berg's technique to overcome this difficulty.

After we had obtained our main result, George Elliott extended his classification to include $X_{n,i} = S^1$.

The results and the techniques used in this paper will also be useful in the classification of simple C^*-algebras arising as inductive limit algebras of the above form with the spaces $X_{n,i}$ being arbitrary 1-dimensional spaces [Li] and higher dimensional spaces [EGL]. We would like to briefly explain the ideas here.

Suppose that A is a simple inductive limit as above with spaces $X_{n,i}$ arbitrary finite CW complexes. If one further supposes that A is of real rank zero (i.e. the set of invertible self adjoint elements in A is dense in the set of self adjoint elements in A), then TA is completely determined by K_0A via the canonical pairing map. In this case, one can prove the following. For any finite set $F \subset A_n$, the homomorphism $\phi_{n,\infty} : A_n \to A$, and any $\varepsilon > 0$, there is a homomorphism $\flat : A_n \to A$ factoring through $\hat{A}_n = \oplus_{i=1}^{k_n} M_{[n,i]}(C(\hat{X}_{n,i}))$, where $\hat{X}_{n,i} \subset X_{n,i}$ are sets of finitely many points, such that

$$||T\phi_{n,\infty}(\tau)(f) - T\psi(\tau)(f)|| \leq \varepsilon$$

for all $\tau \in TA$ and $f \in F$, where the maps $T\phi_{n,\infty}$ and $T\psi$ are induced by $\phi_{n,\infty}$ and ψ, respectively. This fact enables Elliott and Gong to obtain a decomposition result (see 2.21 of [EG]) saying that the homomorphism $\phi_{n,m}$ (for m large enough) can be decomposed into two parts, one with a small supporting projection (for the realization of the map between reduced K-theory) and the other with factorization through a finite dimensional algebra— a direct sum of matrix algebras over a point (for the realization of the map between tracial state spaces). This result plays the key role in the classification of the real rank zero case.

A general simple inductive limit algebra may have real rank **one**, while the above fact (or the decomposition result) is true only if A is of real rank zero (see [BBEK] and [BrE]). On the other hand, for a simple inductive limit algebra, the following analogue is true. For any finite set $F \subset A_n$, the homomorphism $\phi_{n,\infty} : A_n \to A$, and any $\varepsilon > 0$, there is a homomorphism $\psi : A_n \to A$ factoring through

$$\hat{A}_n = \oplus_{i=1}^{k_n} M_{[n,i]}(C(\hat{X}_{n,i})) \ ,$$

where $\hat{X}_{n,i} \subset X_{n,i}$ are trees such that

$$||T\phi_{n,\infty}(\tau)(f) - T\psi(\tau)(f)|| \leq \varepsilon$$

for each $\tau \in A$ and $f \in F$. (In order to do this, the tree $\hat{X}_{n,i}$ should be chosen to be a subset which is dense in $X_{n,i}$ to a certain extent. This can not be done if one replaces the tree by an interval.)

In this case one can only obtain a weaker form of the decomposition results— the homomorphism is decomposed into two parts with its major part (the one with a larger supporting projection) factoring through a matrix algebra over trees (instead of finitely many points). (See [Li] and [EGL] for details.) Therefore, the case of $X_{n,i}$ being trees is a crucial step.

We introduce some notations.

1.1. In this paper, we assume that all the inductive limits $A = \lim_{n \to \infty} (A_n, \phi_{n,m})$ satisfy $A \neq M_k(\mathbb{C})$ for any k. (That is, A is not a matrix algebra.) We also assume that, for any summand A_n^i, $\phi_{n,n+1}(\mathbf{1}_{A_n^i}) \neq 0$, since otherwise, we can simply delete A_n^i from A_n without changing the inductive limit.

With the above assumption, it is well known that for each n (see [BDR]),

$$\lim_{m \to \infty} \min_{\substack{i,j \\ y \in X_{m,j}}} \operatorname{rank}(\phi_{n,m}^{i,j}(\mathbf{1}_{A_n^i})(y)) = +\infty,$$

where the algebra A_n^i is the ith block of A_n, and $\phi_{n,m}^{i,j} : A_n^i \to A_m^j$ is the partial map of the homomorphism $\phi_{n,m}$ from the ith block of A_n to the jth block of A_m.

1.2. For a unital C^*-algebra A, let TA denote the space of tracial states of A, i.e., $\tau \in TA$, if and only if τ is a positive linear map from A to the complex plane $\mathbb{C}$, with $\tau(xy) = \tau(yx)$ and $\tau(\mathbf{1}) = 1$. $AffTA$ is the collection of all the affine maps from TA to $\mathbb{C}$. (In most references, $AffTA$ is defined to be the set of all the affine maps from TA to $\mathbb{R}$. Our $AffTA$ is a complexification of the standard $AffTA$.) An element $x \in AffTA$ is said to be positive if $x(\tau) \geq 0$ for all $\tau \in TA$. And the element $\mathbf{1} \in AffTA$, defined by

$$\mathbf{1}(\tau) = 1$$

for all $\tau \in TA$, is called the unit of $AffTA$. $AffTA$, together with the positive cone $AffTA_+$ and the unit element $\mathbf{1}$ form a scaled ordered complex Banach space. (Notice that for any element $x \in AffTA$, there are $x_1, x_2, x_3, x_4 \in AffTA_+$ such that $x = x_1 - x_2 + ix_3 - ix_4$.)

1.3. For a unital C^*-algebra A, let $V(A)$ denote the collection of all Murray von Neumann equivalence classes of projections in $\bigcup_{n=1}^{\infty} M_n(A)$. Define

$$K_0(A) = \{(a, b); \quad a \in V(A), b \in V(A)\}/ \sim,$$

where $(a, b) \sim (a', b')$ if and only if there is a $c \in V(A)$ such that

$$a + b' + c = a' + b + c \in V(A).$$

(The element (a, b) is also written as $a - b$.)

Let $K_0(A)_+$ be the image of $V(A)$ in $K_0(A)$, i.e.

$$K_0(A)_+ = \{(a, 0) \in K_0(A); \quad a \in V(A)\}.$$

1.4. The pairing $< \cdot, \cdot >: \ TA \times K_0(A) \to \mathbb{R}$ is defined by

$$< \tau, \ [p] - [q] > = \sum_{i=1}^{k} \tau(p_{ii}) - \sum_{i=1}^{k} \tau(q_{ii}),$$

where $[p] - [q] \in K_0(A)$ is represented by the formal difference of two projections $p, q \in M_k(A)$.

1.5. Any unital homomorphism $\phi : A \to B$ induces an affine map

$$AffT\phi : AffTA \longrightarrow AffTB.$$

Suppose that $P \in M_l(C(X))$ is a non-zero projection with constant rank. It is well known that

$$AffT(PM_l(C(X))P) = AffT(M_l(C(X))) = C(X) .$$

If $\phi : C(X) \to M_l(C(Y))$ is a unital homomorphism, then $AffT\phi : C(X) \to C(Y)$ is determined by

$$AffT\phi(f) = \frac{1}{l} \sum_{i=1}^{l} \phi(f)_{ii},$$

where each $\phi(f)_{ii}$ is the entry of $\phi(f) \in M_l(C(Y))$ at the place (i, i).

In general, if $\phi : C(X) \to PM_l(C(Y))P$ is a unital homomorphism and $\mathrm{rank}(P) = k$, then $AffT\phi : C(X) \to C(Y)$ is defined by

$$AffT\phi(f) = \frac{1}{k} \sum_{i=1}^{l} \phi(f)_{ii},$$

where each $\phi(f)_{ii}$ is the entry of $\phi(f) \in PM_l(C(Y))P \subset M_l(C(Y))$ at the place (i,i).

1.6. Let $\phi_1 : C(X) \to PM_{l_1}(C(Y))P$ and $\phi_2 : C(X) \to QM_{l_2}(C(Y))Q$ be two unital homomorphisms. Set

$$\phi = \mathrm{diag}(\phi_1, \phi_2) : C(X) \longrightarrow (P \oplus Q)M_{l_1+l_2}(C(Y))(P \oplus Q).$$

Then by 1.5,

$$AffT\phi = \frac{k_1}{k_1 + k_2} AffT\phi_1 + \frac{k_2}{k_1 + k_2} AffT\phi_2,$$

where $k_1 = \mathrm{rank}(P)$ and $k_2 = \mathrm{rank}(Q)$. Also, if P and Q are orthogonal projections in $M_l(C(Y))$, then $\phi = \mathrm{diag}(\phi_1, \phi_2)$ can be considered to be a homomorphism from $C(X)$ to $(P + Q)M_l(C(Y))(P + Q)$, and the above equality still holds.

1.7. Suppose that $\phi : C(X) \to PM_l(C(Y))P$ is a unital homomorphism with $\mathrm{rank} P = k$. For each $y \in Y$, there is a unitary $u_y \in M_l(\mathbb{C})$ and $x_1(y), x_2(y), \cdots, x_k(y) \in X$ such that

$$\phi(f)(y) = u_y \begin{pmatrix} f(x_1(y)) & & & & & \\ & \ddots & & & & \\ & & f(x_k(y)) & & & \\ & & & 0 & & \\ & & & & \ddots & \\ & & & & & 0 \end{pmatrix} u_y^* .$$

The set $\{x_1(y), x_2(y), \cdots, x_k(y)\}$ is called the spectrum of ϕ at y, and is denoted by $\mathrm{Sp}\phi_y$. (Note that, for the set $\mathrm{Sp}\phi_y$, we always count the multiplicity.) In other words, there are mutually orthogonal rank 1 projections $p_1, p_2, \cdots, p_k$ with $\sum p_i = P$ such that $\phi(f)(y) = \sum_{i=1}^{k} f(x_i(y))p_i$. Grouping all the same $x_i(y)$ together, we can write

$$\phi(f)(y) = \sum_{i=1}^{k_1} f(\lambda_i(y))P_i, \qquad (k_1 \leq k)$$

where, as a set, $\{\lambda_1(y), \lambda_2(y), \cdots, \lambda_{k_1}(y)\} = \{x_1(y), x_2(y), \cdots, x_k(y)\}$. But $\lambda_i(y) \neq \lambda_j(y)$ if $i \neq j$. Furthermore, if $\lambda_i(y)$ has multiplicity m (i.e., $\lambda_i(y)$ appears m times in $\{x_1(y), x_2(y), \cdots, x_k(y)\}$), then $\mathrm{rank}(P_i) = m$. It is convenient to call P_i the spectral projection of ϕ at y with respect to the spectrum $\lambda_i(y)$.

1.8. Any unital homomorphism $\phi : M_k(C(X)) \to PM_l(C(Y))P$ can be identified with $\phi_1 \otimes \mathbf{1}_k$, for a certain identification of $PM_l(C(Y))P$ with $(pM_l(C(Y))p) \otimes M_k$, where $p = \phi(e_{11})$ and

$$\phi_1 = \phi|_{e_{11}M_k(C(X))e_{11}} : \ C(X) \longrightarrow pM_l(C(Y))p.$$

Define $\mathrm{Sp}\phi_y = \mathrm{Sp}(\phi_1)_y$ for each y.

Let

$$\phi : \ \oplus_{i=1}^{q} M_{k_i}(C(X_i)) \longrightarrow \oplus_{j=1}^{t} P_j M_{l_j}(C(Y_j))P_j$$

be a homomorphism and $Y = \coprod Y_j$ be the disjoint union of the spaces $\{Y_i\}_{i=1}^{t}$. For each $y \in Y$, $y \in Y_j$ for some j. The spectrum of the homomorphism ϕ at the point

$y \in Y$ is defined by

$$\mathrm{Sp}\phi_y = \bigcup_{i=1}^{q} \mathrm{Sp}(\phi^{i,j})_y \, ,$$

where the homomorphism

$$\phi^{i,j} : \; M_{k_i}(C(X_i)) \longrightarrow \phi(\mathbf{1}_{k_i})P_j M_{l_j}(C(Y_j))P_j\phi(\mathbf{1}_{k_i})$$

is the partial map of ϕ corresponding to i, j. Note that

$$\mathrm{Sp}\phi_y = \bigcup_{i=1}^{q} \mathrm{Sp}(\phi^{i,j})_y \subset X := \coprod X_i \, .$$

1.9. The following fact will be frequently used: For homomorphisms $\phi : C(X) \to PM_l(C(Y))P$ and $\phi \otimes \mathbf{1}_n : M_n(C(X)) \to M_n(PM_l(C(Y))P)$ with rank$(P) = k$,

$$AffT\phi(f)(y) = \frac{1}{k} \sum_{x_i(y) \in \mathrm{Sp}\phi_y} f(x_i(y))$$

and

$$AffT(\phi \otimes \mathbf{1}_n) = AffT\phi.$$

1.10. For a C^*-algebra $A = \oplus_{i=1}^{t} P_i M_{l_i}(C(Y_i))P_i$, where P_i are non-zero projections with constant rank, we write $\mathrm{Sp}A = \coprod Y_i$.

1.11. In this paper, all trees, graphs and CW complexes are supposed to be connected unless specified otherwise.

1.12. Let X be a metric space. Two k-tuples of (possibly repeating) points $\{x_1, x_2, \cdots, x_k\} \subset X$ and $\{x'_1, x'_2, \cdots, x'_k\} \subset X$ are said to **be paired within** η if there is a permutation σ of $\{1, 2, \cdots, k\}$ such that

$$\text{dist}(x_i, x'_{\sigma(i)}) < \eta, \qquad\qquad i = 1, 2, \cdots, k.$$

1.13. Let X be a compact metric space, $X_1 \subset X$ a closed subset, and $c > 0$ a number. Set

$$B_c(X_1) = \{x;\ \text{dist}(x, X_1) < c\}.$$

Let $X_2 \subset X$ be another subset. We say that $X_2 \subset_c X_1$ if $X_2 \subset B_c(X_1)$.

1.14. Let $|E|$ denote the number of the elements in the set E. When we write $|\text{Sp}\phi_y|$, we count multiplicity.

1.15. In this paper, by a graph, we mean a 1-dimensional simplicial complex — each edge has exactly two vertices and any two edges have at most one intersection point which must be a vertex. For example, S^1 can be regarded as a graph with three edges and three vertices.

Acknowledgment This research was supported by the University of Toronto Fellowships, and the financial assistance of the Department of Mathematics, the University of Toronto. The author would like to take this opportunity to thank G.A. Elliott for the suggestion of the topic and for many helpful discussions.

Chapter 2

Diagonalization, Distinct Spectrum and Injectivity

In this chapter we will prove several results concerning the change of arbitrary homomorphisms to the homomorphisms with certain properties such as diagonalization, distinct spectrum, and injectivity. These properties are to be used later on.

2.1 Diagonalization and distinct spectrum

The ideas used in the proof of diagonalization are essentially contained in §3 of [Su], except Lemma 2.1.1 and Lemma 2.1.2. [Su] deals with a more complicated case which involves multiple points. In this chapter, we will repeat some of his proofs (of the case without multiple points) to make clear how we obtain distinct spectrum (Theorem 2.1.6) and global diagonalization for the case of trees (Theorem 2.1.7). Neither Theorem 2.1.6 nor Theorem 2.1.7 is stated in [Su].

First of all, let us consider a special case. Suppose that $X = X_1 \vee X_2 \vee \cdots \vee X_k$ is a bunch of k intervals $X_i = [0,1]$ $(i = 1, 2, \cdots, k)$ joining at one point as in the figure below.

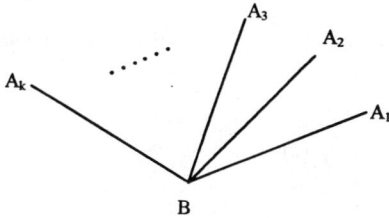

We have the following lemma:

Lemma 2.1.1. *Let $X = X_1 \vee X_2 \vee \cdots \vee X_k$ be a tree described above. Let $Y = [0,1]$. Let $X^0, X^1 \subset X$ be two subsets, each of which contains l distinct points in X.*

It follows that there are l continuous functions $f_1, \cdots, f_l : Y \to X$ such that

(1) As sets, $\{f_1(0), f_2(0), \cdots, f_l(0)\} = X^0$ and $\{f_1(1), f_2(1), \cdots, f_l(1)\} = X^1$.

(2) $f_i(t) \neq f_j(t)$ for any $t \in Y$ and $i \neq j$.

That is, the points in the sets X^0 and X^1 can be connected one by one, by l continuous functions on $[0,1]$ which are distinct at each $t \in [0,1]$.

PROOF: We prove the lemma by induction on k—the number of edges of X.

Let $X^0 = \{x_1^0, x_2^0, \cdots, x_l^0\}$ and $X^1 = \{x_1^1, x_2^1, \cdots, x_l^1\}$. Consider the case

$k = 1$, i.e., $X = [0,1]$ has only one edge. Reorder the sets $\{x_1^0, x_2^0, \cdots, x_l^0\}$ and $\{x_1^1, x_2^1, \cdots, x_l^1\}$ so that

$$x_1^0 < x_2^0 < \cdots < x_l^0 \qquad \text{and} \qquad x_1^1 < x_2^1 < \cdots < x_l^1.$$

For each $i = 1, 2, \cdots, l$, choose a linear function f_i connecting the points x_i^0 and x_i^1. It is obvious that the functions $f_1(t), f_2(t), \cdots, f_l(t)$ are distinct at each point $t \in [0,1]$. Actually

$$f_1(t) < f_2(t) < \cdots < f_l(t)$$

for each $t \in [0,1]$.

Suppose that the lemma is true for $k = n$. Consider a tree X described above with $n + 1$ branches. Set

$$a_i = |X^0 \bigcap [A_i, B]| \qquad \text{and} \qquad b_i = |X^1 \bigcap [A_i, B]|$$

for $1 \le i \le n + 1$. That is, a_i (b_i respectively) is the number of the elements of $X^0 = \{x_j^0\}_{j=1}^l$ ($X^1 = \{x_j^1\}_{j=1}^l$ respectively) which lies inside the ith branch (not including the point B) of X. (Recall that $|E|$ is the cardinal number of the set E.)

Let $1 \le i_0 \le k$ be a number such that

$$|a_{i_0} - b_{i_0}| = \min_{1 \le j \le n+1} |a_j - b_j|.$$

If $a_{i_0} = b_{i_0}$, i.e., $|X^0 \bigcap [A_{i_0}, B]| = |X^1 \bigcap [A_{i_0}, B]|$, then there exist distinct continuous functions

$$f_1, f_2, \cdots, f_{a_{i_0}} : Y \longrightarrow [A_{i_0}, B)$$

connecting all the points in the sets $\{x_j^0\}_{j=1}^l \cap [A_{i_0}, B)$ and $\{x_j^1\}_{j=1}^l \cap [A_{i_0}, B)$, by the same method as used in the case of $k = 1$. (That is, consider $[A_{i_0}, B) = [0, 1)$, and reorder the sets $\{x_j^0\}_{j=1}^l \cap [A_{i_0}, B)$ and $\{x_j^1\}_{j=1}^l \cap [A_{i_0}, B)$. Then connect the corresponding pairs by linear functions.) Furthermore, by induction assumption, there are distinct continuous functions $f_{a_{i_0}+1}, \cdots, f_l : Y \to X'$ connecting $\{x_j^0\}_{j=1}^l \cap X'$ and $\{x_j^1\}_{j=1}^l \cap X'$, where

$$
\begin{aligned}
X' &= X \setminus [A_{i_0}, B) \\
&= [A_1, B] \bigvee \cdots \bigvee [A_{i_0-1}, B] \bigvee [A_{i_0+1}, B] \cdots \bigvee [A_{n+1}, B]
\end{aligned}
$$

has only n branches.

If $a_{i_0} \neq b_{i_0}$, then, without loss of generality, we can assume that $a_{i_0} < b_{i_0}$. (The case of $a_{i_0} > b_{i_0}$ is the same. One can interchange the roles of 0 and 1 in $Y = [0, 1]$.) Since both $\sum_{j=1}^k a_j$ and $\sum_{j=1}^k b_j$ are either l or $l - 1$ (depending on whether the point B is in the set or not), and $a_{i_0} < b_{i_0}$, there is at least one index i_1 such that $a_{i_1} \geq b_{i_1}$. Therefore,

$$a_{i_1} - b_{i_1} \geq b_{i_0} - a_{i_0}.$$

(Notice that $b_{i_0} - a_{i_0}$ is the minimal one among all $|a_i - b_i|$.) Hence

$$a_{i_1} \geq b_{i_0} - a_{i_0}.$$

Identifying $[A_{i_0}, B] \bigcup [B, A_{i_1}]$ with the interval $[0, 2]$ (i.e., identifying $[A_{i_0}, B]$ with

$[0, 1]$, and $[B, A_{i_1}]$ with $[1, 2]$). Let $z_1, z_2, \cdots, z_{b_{i_0}} \in [A_{i_0}, B] \bigcup [B, A_{i_1}] = [0, 2]$ be

the first b_{i_0} numbers in $\{x_j^0\}_{j=1}^l \bigcap ([A_{i_0}, B] \bigcup [B, A_{i_1}])$ from 0 to 2. Notice that

$a_{i_1} \geq b_{i_0} - a_{i_0}$, and $|\{x_j^0\}_{j=1}^l \bigcap ([A_{i_0}, B] \bigcup [B, A_{i_1}])|$ is either equal to $a_{i_0} + a_{i_1}$ or

equal to $a_{i_0} + a_{i_1} + 1$ (depending on whether the point B is in the set or not).

There are at least b_{i_0} elements in the set $\{x_j^0\}_{j=1}^l \bigcap ([A_{i_0}, B] \bigcup [B, A_{i_1}])$.

Let $w_1, w_2, \cdots, w_{b_{i_0}}$ be all the points in $\{x_j^1\}_{j=1}^l \bigcap [A_{i_0}, B)$ in increasing order.

(Recall that $[A_{i_0}, B)$ is identified with $[0, 1)$.)

Define continuous functions

$$f_1, f_2, \cdots, f_{b_{i_0}} : \quad [0, \frac{1}{2}] \longrightarrow [A_{i_0}, B] \bigcup [B, A_{i_1}]$$

which connect $z_1, z_2, \cdots, z_{b_{i_0}}$ and $w_1, w_2, \cdots, w_{b_{i_0}}$ linearly. Let

$$f_{b_{i_0}+1}, \cdots, f_l : [0, \frac{1}{2}] \longrightarrow X$$

be constant functions satisfying

$$\{f_{b_{i_0}+1}(x), \cdots, f_l(x)\} = \{x_i^0\}_{i=1}^l \setminus \{z_i\}_{i=1}^{b_{i_0}}, \qquad x \in [0, \frac{1}{2}] .$$

Obviously, $f_1(t), f_2(t), \cdots, f_l(t)$ are distinct for each $t \in [0, \frac{1}{2}]$. Also, $f_1, f_2, \cdots, f_l$

connect the sets $X^0 = \{x_1^0, x_2^0, \cdots, x_l^0\} \subset X$ and

$$\{x_1^{\frac{1}{2}}, x_2^{\frac{1}{2}}, \cdots, x_l^{\frac{1}{2}}\} := \{f_1(\frac{1}{2}), f_2(\frac{1}{2}), \cdots, f_l(\frac{1}{2})\} .$$

Notice that

$$|\{x_j^{\frac{1}{2}}\}_{j=1}^l \bigcap [A_{i_0}, B)| = |\{x_j^1\}_{j=1}^l \bigcap [A_{i_0}, B)| = b_{i_0}.$$

By induction (see the part of the case $a_{i_0} = b_{i_0}$) one can connect the sets $\{x_j^{\frac{1}{2}}\}_{j=1}^l$ and $\{x_j^1\}_{j=1}^l$ by distinct continuous functions from $[\frac{1}{2}, 1]$ to X.

$$\square$$

Lemma 2.1.2. *Let X be a graph and $X^0 = \{x_1^0, x_2^0, \cdots, x_l^0\} \subset X$ and $X^1 = \{x_1^1, x_2^1, \cdots, x_l^1\} \subset X$ be two subsets as in Lemma 2.1.1. Suppose that $0 < \varepsilon < \dfrac{1}{2}$. And suppose that the sets $X^0 = \{x_i^0\}_{i=1}^l$ and $X^1 = \{x_i^1\}_{i=1}^l$ are paired within $\dfrac{\varepsilon}{6l}$. Then there exist l continuous functions $\alpha_1, \alpha_2, \cdots, \alpha_l : [0, 1] \to X$ such that:*

(1) $\{\alpha_1(0), \alpha_2(0), \cdots, \alpha_l(0)\} = X^0$ and $\{\alpha_1(1), \alpha_2(1), \cdots, \alpha_l(1)\} = X^1$ as sets;

(2) $\alpha_i(t) \neq \alpha_j(t)$ for any $t \in [0, 1]$ and $i \neq j$;

(3) $dist(\alpha_i(t_1), \alpha_i(t_2)) < \varepsilon$ for $1 \leq i \leq l$ and all $t_1, t_2 \in [0, 1]$.

PROOF: Divide $X^0 = \{x_1^0, x_2^0, \cdots, x_l^0\}$ into groups $E_1^0, E_2^0, \cdots, E_{l_1}^0$ as follows. The points x_i^0 and x_j^0 are in the same group if and only if there are $i_0 = i, i_1, \cdots, i_k = j$ such that

$$dist(x_{i_{r-1}}^0, x_{i_r}^0) < \frac{\varepsilon}{2l}, \qquad\qquad r = 1, 2, \cdots, k.$$

This division yields the following two properties:

(a) If x_i^0 and x_j^0 are in the same group, then

$$dist(x_i^0, x_j^0) < \frac{\varepsilon}{2l} \cdot l = \frac{\varepsilon}{2} ;$$

(Notice that each group contains at most l elements.)

(b) If x_i^0 and x_j^0 are in different groups, then

$$\text{dist}(x_i^0, x_j^0) \geq \frac{\varepsilon}{2l} .$$

Let $X_i \subset X$ be the smallest path connected subspace of X containing E_i^0. It is evident that, for each $i \neq j$,

$$\text{dist}(X_i, X_j) \geq \frac{\varepsilon}{2l} ,$$

and for all i,

$$\text{diameter}(X_i) < \frac{\varepsilon}{2} .$$

Also, divide $\{x_1^1, x_2^1, \cdots, x_l^1\}$ into groups $E_1^1, E_2^1, \cdots, E_{l_1}^1$ as follows. Let σ be a permutation of $\{1, 2, \cdots, l\}$ with the property

$$\text{dist}(x_{\sigma(j)}^0, x_j^1) < \frac{\varepsilon}{6l}.$$

E_i^1 is defined to be all those $x_j^1 \in \{x_k^1\}_{k=1}^l$ with $x_{\sigma(j)}^0 \in E_i^0$. Set $E_i = E_i^0 \bigcup E_i^1$ and

$$\tilde{X}_i = \{x \in X; \ \text{dist}(x, X_i) \leq \frac{\varepsilon}{6l}\} .$$

Then $\tilde{X}_i \supset E_i$, and $\tilde{X}_i \bigcap \tilde{X}_j = \varnothing$ if $i \neq j$. Since diameter$(\tilde{X}_i) < \frac{\varepsilon}{2} + \frac{\varepsilon}{6l} < \frac{1}{2}$, $\tilde{X}_i$ are trees of the form described in Lemma 2.1.1. By Lemma 2.1.1, the points in E_i^0 and the points in E_i^1 can be connected by distinct continuous functions within $\tilde{X}_i$.

Notice that (3) follows from

$$\text{diameter}(\tilde{X}_i) < \frac{\varepsilon}{2} + \frac{\varepsilon}{6l} < \varepsilon.$$

The proof is completed.

$\square$

The following two lemmas are contained in [Su].

Lemma 2.1.3. *For any graph X, a finite set of generators $F \subset C(X)$, and $\varepsilon > 0$, there exists $\delta > 0$ such that for any unital homomorphism $\phi : C(X) \to M_n(C(Y))$ and any $y_1, y_2 \in Y$, if*

$$\|\phi(f)(y_1) - \phi(f)(y_2)\| < \delta$$

for all $f \in F$, then $Sp\phi_{y_1}$ and $Sp\phi_{y_2}$ can be paired within ε.

In fact, by Lemma 2.6 in [Su] we know that, if $\phi(h)(y_1)$ and $\phi(h)(y_2)$ are close to within δ up to unitary equivalence for a dense enough finite subset of H, where H is the set of test functions on X, then $Sp\phi_{y_1}$ and $Sp\phi_{y_2}$ can be paired within ε. Since F is a generator set, the condition that $\|\phi(f)(y_1) - \phi(f)(y_2)\|$ is small enough for each $f \in F$ implies that $\|\phi(h)(y_1) - \phi(h)(y_2)\|$ are small for all $h \in H$.

Lemma 2.1.4. *For any graph X, a finite set of generators $F \subset C(X)$, $\varepsilon > 0$ and an integer n, there is a $\delta > 0$ with the following property: If $x_1, x_2, \cdots, x_n \in X$*

are finitely many points and $u, v \in M_n(\mathbb{C})$ are two unitaries such that

$$\left\| u \begin{pmatrix} f(x_1) & & \\ & \ddots & \\ & & f(x_n) \end{pmatrix} u^* - v \begin{pmatrix} f(x_1) & & \\ & \ddots & \\ & & f(x_n) \end{pmatrix} v^* \right\| < \delta$$

for each $f \in F$, then there is a continuous path of unitaries $\{u_t;\ t \in [0,1]\} \subset M_n(\mathbb{C})$ connecting u and v (i.e., $u_0 = u$, $u_1 = v$) such that

$$\left\| u_t \begin{pmatrix} f(x_1) & & \\ & \ddots & \\ & & f(x_n) \end{pmatrix} u_t^* - u_{t'} \begin{pmatrix} f(x_1) & & \\ & \ddots & \\ & & f(x_n) \end{pmatrix} u_{t'}^* \right\| < \varepsilon$$

for each $f \in F$ and $t, t' \in [0,1]$.

This lemma was proved in steps 2 and 3 of the proof of Theorem 3.1 of [Su].

Remark 2.1.5. In this lemma, δ depends not only on the finite set of generators $F \subset C(X)$ and ε, but also on n. Otherwise the lemma is false even for the case $X = S^1$. (For the case of $X = $ tree, one can choose δ to be independent of n. That will be the main result in Chapter 3 of this paper.)

Theorem 2.1.6. *For any graphs X, Y, a finite set of generators $F \subset C(X)$, $\varepsilon > 0$, and a homomorphism $\phi : C(X) \to M_n(C(Y))$, there is a homomorphism $\psi : C(X) \to M_n(C(Y))$ such that:*

(1) $\|\phi(f) - \psi(f)\| < \varepsilon$ for all $f \in F$;

(2) $Sp\psi_y$ are distinct for each $y \in Y$;

(3) For each edge of Y, identifying with $[0,1]$, there are continuous functions $\alpha_1, \alpha_2, \cdots, \alpha_n : [0,1] \to X$ and a unitary $u(t) \in M_n(C[0,1])$ satisfying

$$\psi(f)(t) = u(t) \begin{pmatrix} f(\alpha_1(t)) & & \\ & \ddots & \\ & & f(\alpha_n(t)) \end{pmatrix} u(t)^*$$

for all $t \in [0,1] \subset Y$ and all $f \in C(X)$.

PROOF: The proof is similar to the proof in [Su] (Chapter 3), but we use our Lemma 2.1.2 to make the eigenvalues distinct.

For X, $F \subset C(X)$, $\dfrac{\varepsilon}{4}$ (in the place of ε), and the integer n, there exists $\delta_1 > 0$ (in the place of δ) as in Lemma 2.1.4. Choose $0 < \delta_2 < \delta_1$ such that $\text{dist}(x_1, x_2) < \delta_2$ implies that

$$|f(x_1) - f(x_2)| < \min(\frac{\delta_1}{4}, \varepsilon)$$

for all $f \in F$.

For $F \subset C(X)$, and $\dfrac{\delta_2}{12n} > 0$ (in the place of ε), there exists δ_3 (in the place of δ) as in Lemma 2.1.3. Set $\delta = \min(\varepsilon, \delta_1, \delta_2, \delta_3)$. There exists $\eta > 0$ such that, $\text{dist}(y_1, y_2) \leq \eta$ implies that, for all $f \in F$,

$$\|\phi(f)(y_1) - \phi(f)(y_2)\| < \frac{\delta}{4} . \tag{$*$}$$

Divide Y into small intervals by finitely many dividing points $\{y_1, y_2, \cdots, y_k\} \subset Y$ such that each interval $[y_i, y_{i+1}]$ has a maximum length of η. For convenience, we assume that all the joint points of Y are contained in the set $\{y_1, y_2, \cdots, y_k\}$. For each dividing point y_i, there are n elements $x_1(y_i), x_2(y_i), \cdots, x_n(y_i) \in X$ and a unitary $u_{y_i} \in M_n(\mathbb{C})$ such that

$$\phi(f)(y_i) = u_{y_i} \begin{pmatrix} f(x_1(y_i)) & & \\ & \ddots & \\ & & f(x_n(y_i)) \end{pmatrix} u_{y_i}^*$$

for all $f \in C(X)$.

If y_1, y_2 are adjacent points, then from the above we know that $\{x_j(y_1)\}_{j=1}^n$ and $\{x_j(y_2)\}_{j=1}^n$ can be paired within $\dfrac{\delta_2}{12n}$ (we apply Lemma 2.1.3 and $(*)$).

For each y_i, one can move some points in $\{x_1(y_i), x_2(y_i), \cdots, x_n(y_i)\}$ around within the distance $\dfrac{\delta_2}{24n}$ to make them distinct. We still denote them by $\{x_1(y_i), x_2(y_i), \cdots, x_n(y_i)\}$.

We first define our ψ on each dividing point y_i as follows:

$$\psi(f)(y_i) = u_{y_i} \begin{pmatrix} f(x_1(y_i)) & & \\ & \ddots & \\ & & f(x_n(y_i)) \end{pmatrix} u_{y_i}^*, \qquad (**)$$

where $\{x_1(y_i), x_2(y_i), \cdots, x_n(y_i)\}$ are distinct. (They are the points after moving.)

Obviously, for all $f \in F$,

$$\|\phi(f)(y_i) - \psi(f)(y_i)\| < \frac{\min(\delta_1, \varepsilon)}{4} .$$

It is evident that, for each adjacent pair of points y_1 and y_2, the two sets $\{x_j(y_1)\}_{j=1}^n$ and $\{x_j(y_2)\}_{j=1}^n$ can be paired within $\frac{\delta_2}{6n} = \frac{\delta_2}{12n} + \frac{\delta_2}{24n} + \frac{\delta_2}{24n}$ after moving.

Fix any edge of Y, identifying it with $[0, 1]$. Suppose that $0 = y^0 < y^1 < \cdots < y^m = 1$ are all the dividing points on this edge. We will define ψ on $[0, 1]$ such that the definition of ψ on each y^i coincides with the definition given in $(**)$. Furthermore, ψ is of the form in (3) of our theorem on $[0, 1]$. So we need to define a continuous unitary valued function $u(y)$ $(0 \le y \le 1)$ and continuous functions $\alpha_1, \alpha_2, \cdots, \alpha_n : [0, 1] \to X$. We will define $\{\alpha_i\}_{i=1}^n$ and u on each interval $[y^i, y^{i+1}]$ one by one.

Since the sets $\{x_1(y^0), x_2(y^0), \cdots, x_n(y^0)\}$ and $\{x_1(y^1), x_2(y^1), \cdots, x_n(y^1)\}$ can be paired within $\delta_2/6n$, by Lemma 2.1.2, there are continuous functions

$$\alpha_1, \alpha_2, \cdots, \alpha_n : \ [y^0, \frac{y^0 + y^1}{2}] \longrightarrow X$$

which are distinct at every point $t \in [y^0, \frac{y^0+y^1}{2}]$ such that, for each $i = 1, 2, \cdots, n$, $\alpha_i(y^0) = x_i(y^0)$ and $\alpha_i(\frac{y^0+y^1}{2}) = x_{\sigma^0(i)}(y^1)$ for a certain permutation σ^0 of $\{1, 2, \cdots, n\}$, and such that $\operatorname{dist}(\alpha_i(y'), \alpha_i(y)) < \delta_2$ for any $y', y \in [y^0, \frac{y^0+y^1}{2}]$. Define

$$u(y) = u_{y^0} \qquad\qquad y \in [y^0, \frac{y^0 + y^1}{2}],$$

where u_{y^0} is the unitary (mentioned before) with

$$\psi(f)(y^0) = u_{y^0} \begin{pmatrix} f(x_1(y^0)) & & \\ & \ddots & \\ & & f(x_n(y^0)) \end{pmatrix} u_{y^0}^*$$

(See (∗∗)). Hence the definition of ψ on $[y^0, \frac{y^0+y^1}{2}]$ is given by

$$\psi(f)(y) = u_{y^0} \begin{pmatrix} f(\alpha_1(y)) & & \\ & \ddots & \\ & & f(\alpha_n(y)) \end{pmatrix} u_{y^0}^*$$

for all $y \in [y^0, \frac{y^0+y^1}{2}]$. Since $\mathrm{dist}(\alpha_i(y'), \alpha_i(y)) < \delta_2$ for any $y', y \in [y^0, \frac{y^0+y^1}{2}]$, one has

$$\|\psi(f)(y) - \psi(f)(y^0)\| < \frac{\delta_1}{4}$$

for each $y \in [y^0, \frac{y^0+y^1}{2}]$.

On the other hand, there is a unitary u'_{y_1} such that

$$u'_{y^1} \begin{pmatrix} f(x_{\sigma^0(1)}(y^1)) & & \\ & \ddots & \\ & & f(x_{\sigma^0(n)}(y^1)) \end{pmatrix} (u'_{y^1})^*$$

$$= u_{y^1} \begin{pmatrix} f(x_1(y^1)) & & \\ & \ddots & \\ & & f(x_n(y^1)) \end{pmatrix} u_{y^1}^* \quad (= \psi(f)(y^1)),$$

where u_{y^1} is as in $(**)$, for all $f \in C(X)$. (Actually, u'_{y^1} can be chosen such that $u'^*_{y^1} \cdot u_{y^1}$ is the unitary determined by permutation σ^0.)

Apply Lemma 2.1.4 to $u = u_{y^0}, v = u'_{y^1}$ and $x_i = x_{\sigma^0(i)}(y^1)$, where the inequality needed in Lemma 2.1.4 is guaranteed by

$$\|\psi(f)(\frac{y^0+y^1}{2}) - \psi(f)(y^1)\|$$
$$\leq \frac{\delta_1}{4} + \|\psi(f)(y^0) - \psi(f)(y^1)\|$$
$$\leq \frac{\delta_1}{4} + \frac{\delta_1}{4} + \frac{\delta_1}{4} + \|\phi(f)(y^0) - \phi(f)(y^1)\|$$
$$\leq \delta_1.$$

There exists a unitary valued path $u : [\frac{y^0+y^1}{2}, y^1] \to M_n(\mathbb{C})$ connecting the points $u_{y^0} (= u(\frac{y^0+y^1}{2}))$ and u'_{y^1} such that, for all $f \in F$ and $y, y' \in [\frac{y^0+y^1}{2}, y^1]$,

$$\left\| u(y)\mathcal{F}(f(x_{\sigma^0}(y^1)))u(y)^* - u(y')\mathcal{F}(f(x_{\sigma^0}(y^1)))u(y')^* \right\| \leq \frac{\varepsilon}{4} ,$$

where

$$\mathcal{F}(f(x_{\sigma^0}(y^1))) = \begin{pmatrix} f(x_{\sigma^0(1)}(y^1)) & & \\ & \ddots & \\ & & f(x_{\sigma^0(n)}(y^1)) \end{pmatrix}.$$

Notice that $x_{\sigma^0(i)}(y^1) = \alpha_i(\frac{y^0+y^1}{2})$. Define $\alpha_1, \alpha_2, \cdots, \alpha_n$ on $[\frac{y^0+y^1}{2}, y^1]$ to be constant functions:

$$\alpha_i(y) = x_{\sigma^0(i)}(y^1) = \alpha_i(\frac{y^0+y^1}{2})$$

for all $y \in [\frac{y^0+y^1}{2}, y^1]$, $i = 1, 2, \cdots, n$. Hence ψ on $[\frac{y^0+y^1}{2}, y^1]$ is defined by

$$\psi(f)(y) = u(y) \begin{pmatrix} f(x_{\sigma^0(1)}(y^1)) & & \\ & \ddots & \\ & & f(x_{\sigma^0(n)}(y^1)) \end{pmatrix} u(y)^*$$

for all $y \in [\frac{y^0+y^1}{2}, y^1]$. Apparently,

$$\|\phi(f)(y) - \psi(f)(y)\| < \varepsilon$$

for each $y \in [y^0, y^1]$.

Suppose that we have already had the definitions of $\alpha_1, \alpha_2, \cdots, \alpha_n$ and u on $[y^0, y^k]$ (hence we have the definition of ψ on $[y^0, y^k]$) such that

$$u(y^k) \begin{pmatrix} f(\alpha_1(y^k)) & & \\ & \ddots & \\ & & f(\alpha_n(y^k)) \end{pmatrix} u(y^k)^*$$

coincides with the old definition of ψ at y^k. (See (∗∗).) We will define $\alpha_1, \alpha_2, \cdots, \alpha_n$ and u on $[y^k, y^{k+1}]$.

Since as a set,

$$\{\alpha_1(y^k), \alpha_2(y^k), \cdots, \alpha_n(y^k)\} = \{x_1(y^k), x_2(y^k), \cdots, x_n(y^k)\} \,,$$

where $\{x_1(y^k), x_2(y^k), \cdots, x_n(y^k)\}$ is from (∗∗), and the sets $\{x_1(y^k), x_2(y^k), \cdots, x_n(y^k)\}$ and $\{x_1(y^{k+1}), x_2(y^{k+1}), \cdots, x_n(y^{k+1})\}$ can be paired within $\dfrac{\delta_2}{6n}$, we can define $\alpha_1, \alpha_2, \cdots, \alpha_n$ on $[y^k, \frac{y^k+y^{k+1}}{2}]$ such that the definition of each $\alpha_i(y^k)$ coincides with our old definition of $\alpha_i(y^k)$ and

$$\alpha_i\left(\frac{y^k + y^{k+1}}{2}\right) = x_{\sigma^k(i)}(y^{k+1})$$

for a certain permutation σ^k of $\{1, 2, \cdots, n\}$. Define $u(y)$ on $[y^k, \frac{y^k+y^{k+1}}{2}]$ to be the constant unitary $u(y) = u(y^k)$. There is a unitary $u'_{y^{k+1}}$ such that

$$u'_{y^{k+1}} \begin{pmatrix} f(x_{\sigma^k(1)}(y^{k+1})) & & \\ & \ddots & \\ & & f(x_{\sigma^k(n)}(y^{k+1})) \end{pmatrix} u'^*_{y^{k+1}}$$

$$= u_{y^{k+1}} \begin{pmatrix} f(x_1(y^{k+1})) & & \\ & \ddots & \\ & & f(x_n(y^{k+1})) \end{pmatrix} u^*_{y^{k+1}} \quad (= \psi(f)(y^{k+1}))$$

for all $f \in C(X)$.

One can define $u(y)$ on $[\frac{y^k+y^{k+1}}{2}, y^{k+1}]$ such that $u(\frac{y^k+y^{k+1}}{2}) = u(y^k)$ (as before) and $u(y^{k+1}) = u'_{y^{k+1}}$, as we did in the case of $[y^0, y^1]$.

Define $\alpha_1, \alpha_2, \cdots, \alpha_n$ on $[\frac{y^k+y^{k+1}}{2}, y^{k+1}]$ to be

$$\alpha_i(y) = \alpha_i\left(\frac{y^k + y^{k+1}}{2}\right) = x_{\sigma^k(i)}(y^{k+1}), \qquad i = 1, 2, \cdots, n .$$

One can check that, ψ, defined by

$$\psi(f)(y) = u(y) \begin{pmatrix} f(\alpha_1(y)) & & \\ & \ddots & \\ & & f(\alpha_n(y)) \end{pmatrix} u(y)^*,$$

where $y \in [y^k, y^{k+1}]$, satisfies

$$\|\phi(f)(y) - \psi(f)(y)\| < \varepsilon$$

for all $y \in [y^k, y^{k+1}]$ and $f \in F$, and that $\psi(f)(y^{k+1})$ coincides with its old definition $(**)$.

In this procedure, ψ can be defined on each edge separately. Obviously (1), (2), (3) all follow.

$\square$

Corollary 2.1.7. *In Theorem 2.1.6, if we further assume that Y is a tree, then the item (3) in 2.1.6 can be changed to*

(3') There exist continuous functions $\alpha_1, \alpha_2, \cdots, \alpha_n : Y \to X$ and a unitary $u \in M_n(C(Y))$ such that

$$\psi(f)(y) = u(y) \begin{pmatrix} f(\alpha_1(y)) & & \\ & \ddots & \\ & & f(\alpha_n(y)) \end{pmatrix} u(y)^*$$

for all $f \in C(X)$ and all $y \in Y$.

PROOF: One can start with an arbitrary extreme point of Y and construct $u, \alpha_1, \cdots, \alpha_n$ piece by piece as we did in Theorem 2.1.6. Note that if we meet a joint point, then we construct $u, \alpha_1, \cdots, \alpha_n$ separately in the coming edges (they only depend on the values of $u, \alpha_1, \cdots, \alpha_n$ at the joint point). Note also that $u, \alpha_1, \cdots, \alpha_n$ are well defined because Y is a tree. We will not have to define $u, \alpha_1, \cdots, \alpha_n$ twice at any point since there is no circle inside Y.

$\square$

Corollary 2.1.8. *Theorem 2.1.6 and Corollary 2.1.7 hold for the more general case*

$$\phi: \ M_k(C(X)) \longrightarrow M_{nk}(C(Y)).$$

In the proof of Theorem 2.1.6, we have also proved the following result which will be used later on.

Corollary 2.1.9. *Let $F \subset C(X)$ be a finite set of generators of $C(X)$. For any $\varepsilon > 0$ and any positive integer n, there exists $\delta > 0$ such that, if homomorphisms $\phi, \psi : C(X) \to M_n(\mathbb{C})$ satisfy $\|\phi(f) - \psi(f)\| < \delta$ for all $f \in F$, then there is a continuous path of homomorphisms $\phi_t : C(X) \to M_n(\mathbb{C})$ satisfying the following conditions:*

(1) $\phi_0 = \phi; \quad \phi_1 = \psi;$

(2) $\|\phi_t(f) - \phi(f)\| < \varepsilon$ for all $f \in F$ and all $t \in [0, 1]$.

2.2   Injectivity

In this section, we will prove the following theorem.

Theorem 2.2.1.   *Let* $A = \lim_{n\to\infty}(A_n, \phi_{n,m})$ *be a unital inductive limit* C^*-*algebra with* $A_n = \oplus_{i=1}^{k_n} M_{[n,i]}(C(X_{n,i}))$, *where* $X_{n,i}$ *are graphs. Then one can write* $A = \lim_{n\to\infty}(B_n, \psi_{n,m})$, *where the algebras* $B_n = \oplus_{i=1}^{l_n} M_{\{n,i\}}(C(Y_{n,i}))$ *are finite direct sums of matrix algebras over graphs* $Y_{n,i}$, *with one extra property that all homomorphisms* $\psi_{n,m}$ *are* **injective**.

For the case $X_{n,i} = S^1$, the theorem was stated in [El] (Theorem 4.3). But the proof was not complete, and the result was not used in [El]. In the proof of $(iii) \Longrightarrow (iv)$, on page 185 of [El], the homomorphisms $A_i \to A_j$ in the diagram

$$
\begin{array}{ccccccc}
A_0 & \longrightarrow & A_1 & \longrightarrow & A_2 & \longrightarrow & \cdots A \\
\downarrow & & \downarrow & & \downarrow & & \\
A & \longrightarrow & A & \longrightarrow & A & \longrightarrow & \cdots A
\end{array}
\quad ,
$$

are not automatically injective, even though the limit is isomorphic to A. (Namely, the map from A_0 to the limit A may not be the same map $A_0 \hookrightarrow A$ in the diagram. They are only approximately the same.)

In this section we will prove that one can choose $A_0 \to A_1 \to A_2 \to \cdots$ in the above diagram to be injective.

The above theorem is used in the proof of classification theory in [El1].

2.2.2. Let $C \subset [0,1]$ be the Cantor set. It is well known that there is a non-decreasing surjective continuous map $\alpha : C \to [0,1]$. (This result can be found in any standard textbook on Real Analysis.)

If $Y \subset [0,1]$ is a closed subset containing uncountably many points, then there is a subspace of Y which is homeomorphic to the space of the Cantor set. Based on this fact, it is easy to prove the following lemma:

Lemma 2.2.3. *Let $Y \subset [0,1]$ be a closed subset containing uncountably many points. Then there exists a surjective non-decreasing continuous map*

$$\alpha : Y \to [0,1].$$

2.2.4. Let $\tilde{Y}$ be a graph, and $Y \subset \tilde{Y}$ a closed subset. We will choose a division of Y, and associate it with a space Z and a continuous surjective map $\alpha : Y \to Z$, where Z is a union of finitely many connected graphs. (Notice that such a surjective map induces an injective homomorphism $\alpha^* : M_n(C(Z)) \to M_n(C(Y))$. Hence it provides a sub-algebra $M_n(C(Z))$ of $M_n(C(Y))$, which is a matrix algebra over finitely many graphs.)

Suppose that $\hat{Y} = \{y_1, y_2, \cdots, \} \subset Y$ is the collection of the finitely many dividing points. Let $V(\tilde{Y})$ denote the set of all the vertices of $\tilde{Y}$. We assume that $\hat{Y} \supset V(\tilde{Y}) \cap Y$, that is, it contains all the vertices of $\tilde{Y}$ which are also in Y. More generally, for any edge of $\tilde{Y}$, identified with $[0,1]$, both the maximum and the minimum of $[0,1] \cap Y \subset \tilde{Y}$ are in $\hat{Y}$.

Two points $y_i, y_j \in \hat{Y}$ are said to be **adjacent** if y_i, y_j are in the same edge of $\tilde{Y}$, and inside the open interval (y_i, y_j), there is no other point in $\hat{Y}$.

It is obvious that Y can be written as the union of $[y_i, y_j] \bigcap Y$, where $[y_i, y_j]$ runs over all adjacent pairs.

We will define a space Z which is a union of finitely many connected graphs, and define a continuous surjective map $\alpha : Y \to Z$ as follows.

First, Z has vertices $V(Z) = \{z_1, z_2, \cdots\}$, each z_i corresponding to one and only one y_i in $\hat{Y} \subset Y$. To define the edges of Z, we consider each interval $[y_i, y_j]$. There are the following two cases.

<u>Case 1.</u> If $[y_i, y_j] \bigcap Y$ has uncountably many points, then we let Z contain $[z_i, z_j]$, the line segment connecting z_i and z_j. We call it $[z_i, z_j]$. By Lemma 2.2.3, there is a non-decreasing surjective map $\alpha : [y_i, y_j] \bigcap Y \to [z_i, z_j]$. (Here both $[y_i, y_j]$ and $[z_i, z_j]$ are identified with interval $[0, 1]$.) Obviously, $\alpha(y_i) = z_i$, and $\alpha(y_j) = z_j$.

<u>Case 2.</u> If $[y_i, y_j] \bigcap Y$ has at most countably many points, then it is defined that there is no edge connecting z_i and z_j. Since $[y_i, y_j] \bigcap Y$ is a countable closed subset of $[y_i, y_j] \subset \tilde{Y}$, there is at least one gap in $[y_i, y_j] \bigcap Y$. That is, there is an open interval $(y_i', y_j') \subset (y_i, y_j)$ (i.e., $y_i' < y_j'$) such that $(y_i', y_j') \bigcap Y = \varnothing$. Let $\alpha : [y_i, y_j] \bigcap Y \to \{z_i, z_j\}$ (the set of two points consisting of z_i, z_j) be defined by

$$\alpha(y) = \begin{cases} z_i & \text{if } y_i \leq y \leq y_i' \\ z_j & \text{if } y_j' \leq y \leq y_j. \end{cases}$$

By the above procedure we obtain a space Z which is a union of finitely many graphs. We also obtain a surjective map $\alpha : Y \to Z$. Notice that α is defined on each $[y_i, y_j] \bigcap Y$ piece by piece, where y_i, y_j are adjacent pairs, and $\alpha(y_i) = z_i$ for each i. The definitions of α on different pieces are consistent.

Certainly, the map α in our construction is not unique. But any such α will work for our purpose.

2.2.5. Let $\eta > 0$. In the above construction, we request the division, with $\widehat{Y} = \{y_1, y_2, \cdots\} \subset Y \subset \tilde{Y}$, to satisfy the following condition: If y_i, y_j are adjacent points and

$(y_i, y_j) \bigcap Y \neq \varnothing$, then $\mathrm{dist}(y_i, y_j) < \eta$.

Lemma 2.2.6. *Let $G \subset M_n(C(Y))$ be a finite subset. Suppose that $\varepsilon > 0$ and $\eta > 0$ satisfy that, $\mathrm{dist}(y, y') \leq \eta$ implies that $\|g(y) - g(y')\| < \varepsilon/2$ for all $g \in G$. Let $\{y_1, y_2, \cdots\}$ be a division of Y satisfying the condition in 2.2.5 for η, and, Z and $\alpha : Y \to Z$ are the space and the surjective map respectively, associated to the above division (as defined in 2.2.4). Then*

$$G \subset_\varepsilon M_n(C(Z)),$$

where $M_n(C(Z))$ is considered to be the sub-algebra of $M_n(C(Y))$ under the inclusion map

$$\alpha^* : M_n(C(Z)) \hookrightarrow M_n(C(Y)).$$

PROOF: For each $g \in G$, define $\tilde{g} \in M_n(C(Z))$ as follows: $\tilde{g}(z_i) = g(y_i)$ for all i; On each interval $[z_i, z_j]$, $\tilde{g}$ is the linear path connecting $\tilde{g}(z_i)$ and $\tilde{g}(z_j)$. Notice that

$$\|\tilde{g}(z) - \tilde{g}(z_i)\| \leq \|\tilde{g}(z_j) - \tilde{g}(z_i)\| = \|g(y_j) - g(y_i)\| < \frac{\varepsilon}{2}$$

for all $z \in [z_i, z_j]$. Hence if $y \in [y_i, y_j]$, then

$$\|\tilde{g}(\alpha(y)) - g(y)\|$$

$$\leq \quad \|\tilde{g}(\alpha(y)) - \tilde{g}(z_i)\| + \|\tilde{g}(z_i) - g(y_i)\| + \|g(y_i) - g(y)\|$$

$$< \quad \frac{\varepsilon}{2} + 0 + \frac{\varepsilon}{2} = \varepsilon$$

That is,

$$\|g - \alpha^*(\tilde{g})\| < \varepsilon$$

as desired.

$\square$

The following lemma is well known (see the proof of Lemma 2.3 of [Br]).

Lemma 2.2.7. *Let $A = \oplus_{k=1}^q M_{l_k}(\mathbb{C})$ and $B = M_n(\mathbb{C})$. Given $\varepsilon > 0$, there exists $\delta > 0$ such that, if $\phi, \psi : A \to B$ satisfy*

$$\|\phi(e_{ij}^k) - \psi(e_{ij}^k)\| < \delta$$

for all matrix units $e_{ij}^k \in M_{l_k}(\mathbb{C})$, then there exists a unitary $u \in M_n(\mathbb{C})$ possessing the following two properties:

(1) $\phi = Adu \circ \psi$;

(2) $\|u - \mathbf{1}\| < \varepsilon$.

2.2.8. Suppose that $\phi, \psi : A = \oplus_{k=1}^q M_{l_k}(C(X_k)) \to B = M_n(\mathbb{C})$ are two unital homomorphisms satisfying $\phi(e_{ij}^k) = \psi(e_{ij}^k)$ for all matrix units e_{ij}^k. Set

$P_k = \phi(\mathbf{1}_{A^k})$ and $p_k = \phi(e_{11}^k)$, where $A^k = M_{l_k}(C(X_k))$ is the kth block of A. There are identifications of $P_k M_n(\mathbb{C}) P_k$ with $M_{l_k}(p_k M_n(\mathbb{C}) p_k)$, $k = 1, 2, \cdots, q$, such that

$$\phi = \operatorname{diag}(\phi_1 \otimes \mathbf{1}_{l_1}, \phi_2 \otimes \mathbf{1}_{l_2}, \cdots, \phi_q \otimes \mathbf{1}_{l_q})$$

$$\psi = \operatorname{diag}(\psi_1 \otimes \mathbf{1}_{l_1}, \psi_2 \otimes \mathbf{1}_{l_2}, \cdots, \psi_q \otimes \mathbf{1}_{l_q}),$$

where $\phi_k, \psi_k : C(X) \to p_k M_n(\mathbb{C}) p_k$ are defined to be the restrictions $\phi|_{e_{11}^k M_{l_k}(C(X_k)) e_{11}^k}$

and $\psi|_{e_{11}^k M_{l_k}(C(X_k)) e_{11}^k}$ respectively. Combining Lemma 2.2.7, 2.2.8 and Corollary 2.1.9, we have the following:

Lemma 2.2.9. *Let $A = \oplus_{k=1}^q M_{l_k}(C(X_k))$, where the spaces X_k are connected graphs. Let $F \subset A$ be a finite set containing all the matrix units e_{ij}^k and containing a generator set of the center $C(X_k)$ of every block $A^k = M_{l_k}(C(X_k))$. For any $\varepsilon > 0$ and any positive integer n, there exists $\delta > 0$ such that, if $\phi, \psi : A \to M_n(\mathbb{C})$ are homomorphisms satisfying $\|\phi(f) - \psi(f)\| < \delta$ for all $f \in F$, then there is a continuous path of homomorphisms $\phi_t : A \to M_n(\mathbb{C})$ with the following properties:*

(1) $\phi_0 = \phi$, $\phi_1 = \psi$;

(2) $\|\phi_t(f) - \phi(f)\| < \varepsilon$ for all $t \in [0,1]$ and all $f \in F$.

The proof of Lemma 2.2.9 is straightforward. One may first connect ϕ with $\operatorname{Ad} u \circ \psi$, then connect $\operatorname{Ad} u \circ \psi$ with ψ, where u is as in Lemma 2.2.7.

Theorem 2.2.10. *Let $A = \oplus_{k=1}^q M_{l_k}(C(X_k))$ and $F \subset A$ be the algebra and* the finite subset given in Lemma 2.2.9. Let $B = \oplus_{i=1}^m M_{n_i}(C(Y_i))$, where each Y_i is

a closed subset of a graph $\tilde{Y}_i$, and let $G \subset B$ be a finite subset. Let $\phi : A \to B$ be an

injective homomorphism. It follows that, for any $\varepsilon > 0$, there exist a sub-algebra

$B' = \oplus_{i=1}^m M_{n_i}(C(Z_i)) \subset B$, where each space Z_i is a union of finitely many graphs,

and an injective homomorphism $\psi : A \to B'$, such that:

(1) $\|\phi(f) - \psi(f)\| < \varepsilon$ for all $f \in F$;

(2) $G \subset_\varepsilon B'$.

PROOF: For $\varepsilon/2 > 0$ and for all the positive integers n_i (in place of n), $i =$

$1, 2, \cdots, m$, there is a common positive number $\delta < 1/4$ (works for all n_i) as

in Lemma 2.2.9. Furthermore, we may require δ to satisfy the condition that

$\mathrm{dist}(x, x') \leq 2\delta$ implies that $\|f(x) - f(x')\| < \varepsilon/2$ for all $f \in F \subset A$. For the

above $\delta > 0$, there exists $\eta > 0$ such that $\mathrm{dist}(y, y') < \eta$ implies that

$$\|\phi(f)(y) - \phi(f)(y')\| < \delta \qquad\qquad \text{for all } f \in F$$

and implies that

$$\|g(y) - g(y')\| < \frac{\varepsilon}{2} \qquad\qquad \text{for all } g \in G.$$

(The latter is the condition in Lemma 2.2.6.)

By Lemma 2.1.3, we may require the above η to satisfy the following condition:

If $\mathrm{dist}(y, y') < \eta$, then $\mathrm{Sp}\phi_y$ and $\mathrm{Sp}\phi_{y'}$ can be paired within δ, where $\mathrm{Sp}\phi_y$ is a

subset of $X = X_1 \coprod X_2 \coprod \cdots \coprod X_q$. (Here $\coprod$ means disjoint union.)

Let Z_i and $\alpha_i : Y_i \to Z_i$ be the spaces and surjective maps defined in 2.2.4

which are associated to a division satisfying the condition in 2.2.5 for the above η. By Lemma 2.2.6

$$G \subset_\varepsilon B' \ (= \bigoplus_{i=1}^{m} M_{n_i}(C(Z_i))) \subset B.$$

We will define an injective homomorphism $\psi : A \to B'$ as follows.

Define ψ on each block $B_i' = M_{n_i}(C(Z_i))$ separately. Fix a block B_i'. Let $z_1, z_2, \cdots\} \subset Z_i$ be vertices of Z_i described in 2.2.4 (corresponding to the division $y_1, y_2, \cdots\} \subset Y_i$). Define

$$\psi(f)(z_k) = \phi(f)(y_k), \qquad z_k \in \{z_1, z_2, \cdots\}.$$

(Note that $\alpha(y_k) = z_k$.)

For each adjacent pair z_k, z_l, connected by the interval $[z_k, z_l]$, define ψ on $[z_k, z_l]$ as follows. For any $y_k \in Y_i$, there exist $x_1, x_2, \cdots, x_s \in X_1 \coprod X_2 \coprod \cdots \coprod X_q$ and a unitary $u \in M_{n_i}(\mathbb{C})$ such that

$$\phi(f)(y_k) = u \begin{pmatrix} f(x_1) & & & \\ & f(x_2) & & \\ & & \ddots & \\ & & & f(x_s) \end{pmatrix} u^*$$

for all $f \in \oplus_{k=1}^{q} M_{l_k}(C(X_k)) = A$.

For each $x_j, 1 \le j \le s$, define a map

$$\beta_j : [z_k, \frac{z_k + z_l}{2}] \longrightarrow X = X_1 \coprod X_2 \coprod \cdots \coprod X_q$$

with the following properties:

(i) $\beta_j(z_k) = x_j$, and $\beta_j(\frac{z_k+z_l}{2}) = x_j$;

(ii) $\mathrm{Im}\beta_j = \overline{B_{2\delta}(x_j)} = \{x \in X;\ \mathrm{dist}(x, x_j) \leq 2\delta\}$ as a set, where $\mathrm{Im}\beta_j = \beta_j([z_k,\ \frac{z_k+z_l}{2}])$ is the image of the elements in $[z_k,\ \frac{z_k+z_l}{2}]$ under β_j. (Note that for two points from two different connected components of X, we always assume the distance between them to be one. Therefore, $\overline{B_{2\delta}(x_j)}$ is a path connected set.)

We define ψ on $[z_k,\ \frac{z_k+z_l}{2}]$ by

$$\psi(f)(z) = u \begin{pmatrix} f(\beta_1(z)) & & & \\ & f(\beta_2(z)) & & \\ & & \ddots & \\ & & & f(\beta_s(z)) \end{pmatrix} u^*.$$

By (i),

$$\psi(f)(\frac{z_k + z_l}{2}) = \psi(f)(z_k) = \phi(f)(y_k)\ .$$

Notice that, from the construction of Z_i in 2.2.4, $\mathrm{dist}(y_k, y_l) < \eta$ in this case. Therefore,

$$\|\psi(f)(z_l) - \psi(f)(\frac{z_k + z_l}{2})\| \ = \ \|\phi(f)(y_l) - \phi(f)(y_k)\| < \delta$$

for all $f \in F$. By Lemma 2.2.9, one can define ψ on $[\frac{z_k+z_l}{2}, z_l]$ so that the definitions of ψ at the points $\frac{z_k+z_l}{2}$ and z_l agree with the ones given before. Furthermore,

$$\|\psi(f)(z) - \psi(f)(\frac{z_k + z_l}{2})\| < \frac{\varepsilon}{2}$$

for all $f \in F$ and $z \in [\frac{z_k + z_l}{2}, z_l]$. Since $\|\beta_j(z) - x_j\| \leq 2\delta$ (See (ii).), we also have

$$\|\psi(f)(z) - \psi(f)(z_k)\| < \frac{\varepsilon}{2}$$

for all $f \in F$ and $z \in [z_k, \frac{z_k + z_l}{2}]$. Thus we have proved the property (1) of our theorem.

We must prove that ψ is injective. We will use the property (ii) of β_j again. Note that the injectivity of ϕ is equivalent to the condition $\bigcup_{y \in \coprod_{i=1}^m Y_i} \mathrm{Sp}\phi_y = X$.

Fix any point $x_0 \in X$, there are two cases.

<u>Case 1.</u> The single point set $\{x_0\}$ is a component of X.

There exists $y \in Y = \coprod_{i=1}^m Y_i$ such that $x_0 \in \mathrm{Sp}\phi_y$. Let $y \in [y_k, y_l]$, where $\{y_k, y_l\}$ is an adjacent pair in $\hat{Y} = \{y_1, y_2, \cdots\}$. Then $\mathrm{Sp}\phi_y$ and $\mathrm{Sp}\phi_{y_k}$ (or $\mathrm{Sp}\phi_{y_l}$) can be paired within $\delta < 1/4$. Therefore, $x_0 \in \mathrm{Sp}\phi_{y_k}$ or $(x_0 \in \mathrm{Sp}\phi_{y_l})$. Notice that there is no other point of X in the $\frac{1}{2}$-neighborhood of x_0. Hence $x_0 \in \mathrm{Sp}\psi_{z_k} = \mathrm{Sp}\phi_{y_k} \subset \bigcup_{y \in Y} \mathrm{Sp}\phi_y$ (or $x_0 \in \mathrm{Sp}\psi_{z_l} = \mathrm{Sp}\phi_{y_l} \subset \bigcup_{y \in Y} \mathrm{Sp}\phi_y$).

<u>Case 2.</u> $x_0 \in X_j$, and X_j is a component of X containing more than one point (hence it contains uncountably many points).

Consider the set $\overline{B_\delta(x_0)} = \{x \in X_j;\ \mathrm{dist}(x, x_0) \leq \delta\}$. Consider also the set $\mathrm{Sp}\phi_{[y_k, y_l]} \cap \overline{B_\delta(x_0)}$ for each interval $[y_k, y_l]$, where $\{y_k, y_l\}$ is an adjacent pair, and $\mathrm{Sp}\phi_{[y_k, y_l]} = \bigcup_{y \in [y_k, y_l] \cap Y} \mathrm{Sp}\phi_y$. Since the union of all such sets $\mathrm{Sp}\phi_{[y_k, y_l]} \cap \overline{B_\delta(x_0)}$ is equal to $\overline{B_\delta(x_0)}$, there is at least one interval $[y_k, y_l]$ such that $\mathrm{Sp}\phi_{[y_k, y_l]} \cap \overline{B_\delta(x_0)}$

is uncountable. Let $[y_k, y_l] \bigcap Y$ be uncountable. Let $x_1 \in \mathrm{Sp}\phi_{[y_k, y_l]} \bigcap \overline{B_\delta(x_0)}$. That

is, $x_1 \in \mathrm{Sp}\phi_y$ for some $y \in [y_k, y_l]$. Since $\mathrm{Sp}\phi_y$ and $\mathrm{Sp}\phi_{y_k}$ can be paired within δ,

there is a point $x_2 \in \mathrm{Sp}\phi_{y_k}$ such that $\mathrm{dist}(x_1, x_2) < \delta$.

By (ii), we know that

$$\mathrm{Sp}\psi_{[z_k, z_l]} \supseteq \overline{B_{2\delta}(x_2)}.$$

But

$$\mathrm{dist}(x_0, x_2) \leq \mathrm{dist}(x_0, x_1) + \mathrm{dist}(x_1, x_2) \leq \delta + \delta = 2\delta.$$

Hence $x_0 \in \mathrm{Sp}\psi_{[z_k, z_l]}$. This ends the proof of the injectivity of ψ.

$\square$

Remark 2.2.11. In the hypothesis of Theorem 2.2.10, we assume that $F \subset A$

is a generator set (containing all the matrix units) as in Lemma 2.2.9. This is only

for convenience since such restriction can be removed — one can always enlarge F

to make it contain a generator set and contain all the matrix units.

2.2.12. Proof of Theorem 2.2.1. Let $\tilde{A}_n = \phi_{n,\infty}(A_n)$, $n = 1, 2, \cdots$. Then

$\tilde{A}_n$ can be expressed as

$$\tilde{A}_n = \oplus_{i=1}^{k_n} M_{[n,i]}(C(\tilde{X}_{n,i})) ,$$

where the spaces $\tilde{X}_{n,i}$ are closed subspaces of the graphs $X_{n,i}$. Write

$A = \lim\limits_{n \to \infty}(\tilde{A}_n, \tilde{\phi}_{n,m})$, where the homomorphisms $\tilde{\phi}_{n,m}$ are induced by $\phi_{n,m}$, and

they are injective.

Let

$$\varepsilon_1 > \varepsilon_2 > \cdots \qquad \text{and} \qquad \eta_1 > \eta_2 > \cdots$$

be two sequences of positive numbers satisfying

$$\sum \varepsilon_n < +\infty \qquad \text{and} \qquad \sum \eta_n < +\infty.$$

Let $\{x_i\}_{i=1}^{\infty}$ be a dense subset of A. We will construct an injective inductive limit $B_1 \to B_2 \to \cdots$ as follows.

Consider $G_1 = \{x_1\} \subset A$. There is an $\tilde{A}_{i_1}$, and a finite subset $\tilde{G}_1 \subset \tilde{A}_{i_1}$ such that $G_1 \subset_{\frac{\varepsilon_1}{2}} \tilde{G}_1$.

For $\tilde{G}_1 \subset \tilde{A}_{i_1}$, applying Lemma 2.2.6, there exists a sub-algebra $B_1 \subset \tilde{A}_{i_1}$ satisfying the following two conditions: (1) B_1 is a finite direct sum of matrix algebras over graphs; (2) $\tilde{G}_1 \subset_{\frac{\varepsilon_1}{2}} B_1$. This gives us the first piece of the diagram

$$\tilde{A}_{i_1}$$
$$\uparrow$$
$$B_1.$$

Let $\{b_{1j}\}_{j=1}^{\infty}$ be a dense subset of B_1. Set $F_1 = \{b_{11}\} \subset B_1$ and $G_2 = \{x_1, x_2\} \subset A$. There exist $\tilde{A}_{i_2}$, $(i_2 > i_1)$ and a finite subset $\tilde{G}_2 \subset \tilde{A}_{i_2}$ such that $G_2 \subset_{\frac{\varepsilon_2}{2}} \tilde{G}_2$. Applying Theorem 2.2.10 to $F_1 \subset B_1, \tilde{G}_2 \subset \tilde{A}_{i_2}$, and the injective map

$$B_1 \hookrightarrow \tilde{A}_{i_1} \to \tilde{A}_{i_2},$$

there exists a sub-algebra $B_2 \hookrightarrow \tilde{A}_{i_2}$, which is a direct sum of matrix algebras over graphs, and an injective map $\psi_{1,2} : B_1 \to B_2$ such that $\tilde{G}_2 \subset_{\frac{\varepsilon_2}{2}} B_2$ and that the following diagram

$$
\begin{array}{ccc}
\tilde{A}_{i_1} & \xrightarrow{\tilde{\phi}_{i_1,i_2}} & \tilde{A}_{i_2} \\
\uparrow & & \uparrow \\
B_1 & \xrightarrow{\psi_{1,2}} & B_2
\end{array}
$$

almost commutes on F_1 to within η_1. Let $\{b_{2j}\}_{j=1}^{\infty}$ be a dense subset of B_2. Choose

$$
F_2 = \{b_{21}, b_{22}\} \bigcup \{\psi_{1,2}(b_{11}), \psi_{1,2}(b_{12})\} \qquad \text{and} \qquad G_3 = \{x_1, x_2, x_3\}
$$

in the places of F_1 and G_2 respectively, and repeat the above construction to obtain $\tilde{A}_{i_3}, B_3 \subset \tilde{A}_{i_3}$ and an injective map $\psi_{2,3} : B_2 \to B_3$. (Use ε_3 in the place of ε_2, and η_2 in the place of η_1.)

In general, we can construct the diagram

$$
\begin{array}{ccccccccc}
\tilde{A}_{i_1} & \xrightarrow{\tilde{\phi}_{i_1 i_2}} & \tilde{A}_{i_2} & \xrightarrow{\tilde{\phi}_{i_2 i_3}} & \tilde{A}_{i_3} & \longrightarrow & \cdots & \tilde{A}_{i_k} & \longrightarrow & \cdots \\
\uparrow & & \uparrow & & \uparrow & & & \uparrow & & \\
B_1 & \xrightarrow{\psi_{1,2}} & B_2 & \xrightarrow{\psi_{2,3}} & B_3 & \longrightarrow & \cdots & B_k & \longrightarrow & \cdots
\end{array}
$$

with the following properties:

(i) The homomorphisms $\psi_{k,k+1}$ are injective;

(ii) For each k, $G_k = \{x_1, x_2, \cdots, x_k\} \subset_{\varepsilon_k} \tilde{\phi}_{i_k,\infty}(B_k)$, where B_k is considered to be a sub-algebra of $\tilde{A}_{i_k}$;

(iii) The diagram

$$
\begin{array}{ccc}
\tilde{A}_{i_k} & \stackrel{\tilde{\phi}_{i_k, i_{k+1}}}{\longrightarrow} & \tilde{A}_{i_{k+1}} \\
\big\uparrow & & \big\uparrow \\
B_k & \stackrel{\psi_{k,k+1}}{\longrightarrow} & B_{k+1}
\end{array}
$$

almost commutes on $F_k = \{b_{ij};\ \ 1 \le i \le k,\ 1 \le j \le k\}$ to within η_k, where $\{b_{ij}\}_{j=1}^{\infty}$ is a dense subset of B_i.

Then, by 2.3 and 2.4 of [El], the above diagram defines a homomorphism from $B = \lim_{n \to \infty}(B_n, \psi_{n,m})$ to $A = \lim_{n \to \infty}(\tilde{A}_n, \tilde{\phi}_{n,m})$. It is routine to check that the homomorphism is in fact an isomorphism. This ends the proof of Theorem 2.2.1.

Chapter 3

Berg Technique

In this chapter we will apply Berg technique to prove Theorem 3.5 and Corollary 3.6 which will be used later on.

3.1. Consider a tree X as a subset of $\mathbb{C}$, the space of complex numbers. Let $V(X)$ be the set of vertices of X and let $E(X)$ be all the edges of X. A vertex of X is called an **extreme point** if it belongs to only one edge of X. If a vertex of X is not an extreme point, then it is called a **joint point**. Identify each edge of X with interval $[0, 1]$, as usual, when we are working inside an edge.

In this chapter, we will always assume the tree X to be embedded into the complex plane $\mathbb{C}$ such that each edge of it is a line segment of length 1. Let h denote the element given by the embedding which is a generator of $C(X)$.

For any two elements $x_1, x_2 \in X \subset \mathbb{C}$, the meaning of the distance between them as complex numbers in $\mathbb{C}$ is clear, denoted by $\mathrm{dist}(x_1, x_2)$, or $|x_1 - x_2|$ as usual. We also need to consider the distance between these two elements along the tree X, which is defined to be the length of the shortest path along X connecting those two points. We denote it by $\text{dist}_X(x_1, x_2)$. Given two edges $e_1, e_2 \in E(X)$, as subsets of the complex plane $\mathbb{C}$, the distance between them is defined be

$$\text{dist}(e_1, e_2) = \inf\{\text{dist}(x_1, x_2); \; x_1 \in e_1, x_2 \in e_2\}.$$

So, if $e_1 \cap e_2 = \varnothing$, then $\text{dist}(e_1, e_2) > 0$. Let d_X denote the smallest one of these numbers. That is,

$$d_X = \min\{\text{dist}(e_1, e_2); \; e_1, e_2 \in E(X) \text{ and } e_1 \cap e_2 = \varnothing\}.$$

Clearly, $d_X > 0$ since X has only finitely many edges.

For a fixed tree $X \subset \mathbb{C}$, if several edges of X joining together at one point, then the angles between any consecutive pair of these edges are also fixed. All such angles form a finite set since X has only finitely many edges. Let θ_X denote the smallest one of these angles. ($\theta_X > 0$.)

The following lemma will be used in the proof of Theorem 3.3. For the case $X = [0, 1]$, the lemma is trivial (one can take $N = 1$).

Lemma 3.2. *Let $X \subset \mathbb{C}$ be a tree, and $N > 0$ and $M > 0$ be integers satisfying $\frac{1}{N} < \min\{\frac{\sin \theta_X}{20}, d_X \cdot \sin \theta_X\}$ and $\frac{1}{M} < \min\{\frac{d_X}{2}, \frac{1}{16}\}$. Let diagonal matrix*

A be defined as follows:

$$A = \begin{pmatrix} \lambda_1 & & & & \\ & \lambda_2 & & & \\ & & \ddots & & \\ & & & & \lambda_n \end{pmatrix} \qquad \lambda_i \in X, \ i = 1, 2, \cdots, n.$$

For the above M and N, if $|\lambda_i - \lambda_{i+1}| < \frac{1}{M^2 N}$ for all $i = 1, 2, \cdots, n$, (define $\lambda_0 = \lambda_n$) then there exists a diagonal matrix

$$B = \mu_1 I_{l_1} \oplus \mu_2 I_{l_2} \oplus \cdots \oplus \mu_r I_{l_r}$$

satisfying the following conditions:

(i) $\|B - A\| < \frac{2}{M}$;

(ii) For each $i = 1, 2, \cdots, r$, $l_i \geq M$, the consecutive pair (μ_{i-1}, μ_i) are on the same edge, and $|\mu_{i-1} - \mu_i| = \frac{1}{M}$, $\mu_1 = \mu_r$.

PROOF: Set $n = pM + r$, $(0 \leq r < M)$. Divide $\{\lambda_1, \lambda_2, \cdots, \lambda_n\}$ into p sets as follows:

$$\begin{aligned} \Lambda_1 &= \{\lambda_1, \cdots, \lambda_M\} \\ \Lambda_2 &= \{\lambda_{M+1}, \cdots, \lambda_{2M}\} \\ &\vdots \\ \Lambda_p &= \{\lambda_{(p-1)M+1}, \cdots, \lambda_{pM+r}\} \ . \end{aligned}$$

Then the number of the elements of each Λ_i is less than $2M$ and not less than M.

For each edge $[A, B]$ of X, identified with $[0, 1]$, let

$$W_{[A,B]} = \{x = \frac{k}{M} \in [A, B];\ k = 0, 1, \cdots, M\}$$

and set $W_X = \bigcup W_{[A,B]}$, where the union is over all the edges of X.

For each Λ_i, choose $\pi(\Lambda_i) \in W_X$ satisfying

$$\text{dist}(\Lambda_i, \pi(\Lambda_i)) = \min_{y \in W_X} \text{dist}(\Lambda_i, y) = \text{dist}(\Lambda_i, W_X).$$

This yields a map $\pi :\ \{\Lambda_1, \Lambda_2, \cdots, \Lambda_p\} \to W_X$. One can prove the following properties:

(1) Each Λ_i has diameter less than $\frac{2M}{NM^2} < \frac{2}{20M} = \frac{1}{10M}$;

(2) For each i and each $\lambda \in \Lambda_i$,

$$\text{dist}(\lambda, \pi(\Lambda_i)) \le \frac{1}{10M} + \frac{1}{2M} < \frac{1}{M}\ ;$$

(3) For each i, the pair $\pi(\Lambda_i), \pi(\Lambda_{i+1})$ are in the same edge of X (hence they are comparable) and

$$|\pi(\Lambda_i) - \pi(\Lambda_{i+1})| \le \frac{1}{M}$$

$$|\pi(\Lambda_p) - \pi(\Lambda_1)| \le \frac{1}{M}\ .$$

The property (3) is proved as follows. For convenience, let

$$\Lambda_i = \{\lambda_1, \cdots, \lambda_M\}, \quad \Lambda_{i+1} = \{\lambda_{M+1}, \cdots, \lambda_{2M}\}.$$

If the set Λ_i is not contained in one edge of X, then there exists j, $1 \leq j < M$ such that λ_j, λ_{j+1} belong to two different edges of X. By assumption $|\lambda_j - \lambda_{j+1}| < \frac{1}{NM^2} < d_X$, the two edges containing λ_j and λ_{j+1}, respectively, have a common vertex, say, J. $\pi(\Lambda_i) = J$ since

$$\text{dist}(\lambda_j, J) \leq \frac{\text{dist}(\lambda_j, \lambda_{j+1})}{\sin \theta_X} \leq \frac{1}{NM^2 \sin \theta_X} < \min\{\frac{1}{10M^2}, \frac{d_X}{M^2}\}$$

and $\text{diameter}(\Lambda_i) \leq \frac{1}{10M}$. ($M$ is large enough so that $\frac{1}{M} < \frac{d_X}{2}$.)

Similarly, since

$$\begin{aligned}
&\text{diameter}(\Lambda_i \bigcup \Lambda_{i+1}) \\
&\leq \quad \text{diameter}(\Lambda_i) + \text{diameter}(\Lambda_{i+1}) + \text{dist}(\Lambda_i, \Lambda_{i+1}) \\
&\leq \quad \frac{1}{10M} + \frac{1}{10M} + \frac{1}{NM^2} \\
&< \quad \frac{3}{10M} ,
\end{aligned}$$

one can prove that, if the set $\Lambda_i \bigcup \Lambda_{i+1}$ is not contained in one edge of X, then those edges intersecting with $\Lambda_i \bigcup \Lambda_{i+1}$ must have one common vertex J, and in this case, $\pi(\Lambda_i) = \pi(\Lambda_{i+1}) = J$.

If Λ_i is contained in one edge of X, then from the property of M we know that, $\pi(\Lambda_i) = J$ belongs to that same edge.

Now consider the case that the two sets Λ_i and Λ_{i+1} lie within one edge of X.

Notice that $\mathrm{dist}(\Lambda_j, \pi(\Lambda_j)) \le \frac{1}{2M}$. We have

$$\mathrm{dist}(\pi(\Lambda_i), \pi(\Lambda_{i+1}))$$

$$\le \quad \mathrm{diameter}(\Lambda_i) + \mathrm{diameter}(\Lambda_{i+1}) + \mathrm{dist}(\Lambda_i, \Lambda_{i+1}) +$$

$$+\mathrm{dist}(\pi(\Lambda_i), \Lambda_i) + \mathrm{dist}(\pi(\Lambda_{i+1}), \Lambda_{i+1})$$

$$\le \quad \frac{1}{10M} + \frac{1}{10M} + \frac{1}{NM^2} + \frac{1}{2M} + \frac{1}{2M}$$

$$< \quad \frac{3}{2M} < \frac{2}{M} \,.$$

But $\mathrm{dist}(\pi(\Lambda_i), \pi(\Lambda_{i+1}))$ must be a multiple of $\frac{1}{M}$ since the two elements $\pi(\Lambda_i)$, $\pi(\Lambda_{i+1})$ lie within a same edge of X. So $\mathrm{dist}(\pi(\Lambda_i), \pi(\Lambda_{i+1})) \le \frac{1}{M}$. Hence it must be 0 or $\frac{1}{M}$.

We want to make $\pi(\Lambda_1) = \pi(\Lambda_p)$. So we need to prove

(4) If $\pi(\Lambda_1) \ne \pi(\Lambda_p)$, then either $\pi(\Lambda_2) = \pi(\Lambda_1)$ or $\pi(\Lambda_2) = \pi(\Lambda_p)$.

By (1),

$$\mathrm{diameter}(\Lambda_p \bigcup \Lambda_1 \bigcup \Lambda_2) \le \frac{3}{10M} + \frac{2}{NM^2} < \frac{4}{10M}$$

and $\pi(\Lambda_i) \in B_{\frac{1}{2M}}(\Lambda_i)$, where

$$B_{\frac{1}{2M}}(\Lambda_i) = \{x \in X; \ \mathrm{dist}(x, \Lambda_i) \le \frac{1}{2M}\}.$$

Hence

$$\mathrm{diameter}(B_{\frac{1}{2M}}(\Lambda_p \bigcup \Lambda_1 \bigcup \Lambda_2)) \le \frac{4}{10M} + \frac{1}{2M} + \frac{1}{2M} < \frac{2}{M}.$$

It follows that, if $\pi(\Lambda_2), \pi(\Lambda_1), \pi(\Lambda_p)$ are in the same edge of X, then $\pi(\Lambda_2) \in \{\pi(\Lambda_1), \pi(\Lambda_p)\}$. Hence $\mathrm{dist}(\pi(\Lambda_p), \pi(\Lambda_2)) \le \frac{1}{M}$. We know that Λ_1 and Λ_p are in the same edge of X since $\pi(\Lambda_1) \neq \pi(\Lambda_p)$. As we have already proved, if Λ_1 and Λ_2 are not contained in the same edge of X, then $\pi(\Lambda_1) = \pi(\Lambda_2)$. Therefore, it is always true that $\pi(\Lambda_2) \in \{\pi(\Lambda_1), \pi(\Lambda_p)\}$. Furthermore, for any $\lambda \in \Lambda_1$,

$$\operatorname{dist}(\lambda, \pi(\Lambda_p)) \leq \operatorname{dist}(\lambda, \pi(\Lambda_1)) + \frac{1}{M} < \frac{2}{M}.$$

We can simply change $\pi(\Lambda_1)$ to $\pi(\Lambda_p)$ ($|\pi(\Lambda_1) - \pi(\Lambda_2)| \leq \frac{1}{M}$ is still true).

Now we have a division of $\{\lambda_1, \lambda_2, \cdots, \lambda_n\}$,

$$\{\underbrace{\lambda_1, \lambda_2, \cdots, \lambda_M}_{\Lambda_1}, \underbrace{\lambda_{M+1}, \cdots, \lambda_{2M}}_{\Lambda_2}, \cdots, \underbrace{\lambda_{n-M+1}, \cdots, \lambda_n}_{\Lambda_p}\},$$

where each Λ_i corresponds to an element $\pi(\Lambda_i) \in W_X$ such that, for all $\lambda \in \Lambda_i, \operatorname{dist}(\lambda, \pi(\Lambda_i)) < \frac{2}{M}$. In addition, $\operatorname{dist}(\pi(\Lambda_i), \pi(\Lambda_{i+1})) \leq \frac{1}{M}$ for each i. For any j with $1 \leq j \leq p-1$, if $\pi(\Lambda_j) = \pi(\Lambda_{j+1})$, then we put Λ_j and Λ_{j+1} together in one group. By this way, we can change our division $\{\Lambda_1, \Lambda_2, \cdots, \Lambda_p\}$ to $\{\tilde{\Lambda}_1, \tilde{\Lambda}_2, \cdots, \tilde{\Lambda}_q\}$ (with $q \leq p$) and let $\pi(\tilde{\Lambda}_i) = \pi(\Lambda_j)$ if $\Lambda_j \subset \tilde{\Lambda}_i$.

Let l_i be the number of elements of $\tilde{\Lambda}_i$, $\mu_i = \pi(\tilde{\Lambda}_i)$ and let

$$B = \mu_1 I_{l_1} \oplus \mu_2 I_{l_2} \oplus \cdots \oplus \mu_q I_{l_q}.$$

Then B satisfies the conditions (i) and (ii) in our lemma.

$\square$

Theorem 3.3. *Let X be a tree in the complex plane $\mathbb{C}$. For any $\varepsilon > 0$, there exists $\delta > 0$ such that, if matrices*

$$A = \begin{pmatrix} \lambda_1 & & & \\ & \lambda_2 & & \\ & & \ddots & \\ & & & \lambda_n \end{pmatrix} \quad and \quad u = \begin{pmatrix} 0 & 0 & \cdots & 1 \\ 1 & 0 & \cdots & 0 \\ \vdots & \ddots & \ddots & \vdots \\ 0 & \cdots & 1 & 0 \end{pmatrix},$$

with each $\lambda_i \in X$, satisfy $\|uAu^* - A\| < \delta$, then under a certain basis, there exist a diagonal matrix $B = b_1 I_{l_1} \oplus b_2 I_{l_2} \oplus \cdots \oplus b_q I_{l_q}$, where $b_i \in X$, and a unitary $v \in M_n(\mathbb{C})$, such that

$$\|B - A\| < \varepsilon, \qquad \|u - v\| < \varepsilon, \qquad vBv^* = B,$$

where $b_i \in X$, $i = 1, 2, \cdots, q$.

PROOF: Choose $\delta = \frac{1}{M^2 N}$. (M is to be determined later, but it is at least as large as required in Lemma 3.2, and N is as in Lemma 3.2.) By Lemma 3.2, at the expense of a perturbation of norm at most $2/M$, we may assume that A has constant blocks

$$A = \mu_1 I_{l_1} \oplus \mu_2 I_{l_2} \oplus \cdots \oplus \mu_q I_{l_q}$$

with the μ_i's satisfying the conditions in Lemma 3.2:

(i) $\mu_i \in W_X$, $i = 1, 2, \cdots, q$, (W_X was defined in the proof of Lemma 3.2);

(ii) For each $i = 1, 2, \cdots, q$, $|\mu_i - \mu_{i+1}| \leq \frac{1}{M}$, $l_i \geq M$ and $\mu_q = \mu_1$.

Let $\mu_1 = \mu_q = O$ be at the origin of $\mathbb{C}$. Define a partial order on X as follows.

For $x_1, x_2 \in X$, we say $x_1 < x_2$ if the shortest path from O to x_2 along X passes through x_1. By the construction in the proof of the above lemma, we have:

(iii) Every consecutive pair μ_i and μ_{i+1} are comparable, i.e., either $\mu_i < \mu_{i+1}$ or $\mu_i > \mu_{i+1}$. Hence if μ_i is a relatively maximal one, then $\mu_{i+1} = \mu_{i-1}$ (assuming $\mu_{q+1} = \mu_1$).

Now suppose that b is one of the relative maximal in $\mu_1, \mu_2, \cdots, \mu_q$. Then

$$
A = \begin{pmatrix}
\ddots & & & & & & & & \\
 & a & & & & & & & \\
 & & \ddots & & & & & & \\
 & & & a & & & & & \\
 & & & & b & & & & \\
 & & & & & \ddots & & & \\
 & & & & & & b & & \\
 & & & & & & & a & \\
 & & & & & & & & \ddots \\
 & & & & & & & & & a \\
 & & & & & & & & & & \ddots
\end{pmatrix}.
$$

Using Berg's idea (see [Be]), write the basis as follows:

$$\cdots \phi_{-1}, \phi_0, \phi_1, \cdots, \phi_k, \phi_{k+1}, \cdots, \phi_{r-k}, \phi_{r-k+1}, \cdots, \phi_r, \phi_{r+1}, \cdots ,$$

where $\frac{M}{4} \le k \le \frac{M}{2} - 1$. Let u be a shift, $u\phi_i = \phi_{i+1}$ for all i, and let A satisfy

$A\phi_{-1} = a\phi_{-1}$, $A\phi_i = b\phi_i$, $0 \le i \le r$, $A\phi_{r+1} = a\phi_{r+1}$. Replace the $2k+2$ elements (of the basis)

$$\phi_0, \phi_1, \cdots, \phi_k, \phi_{r-k}, \phi_{r-k+1}, \cdots, \phi_r$$

by

$$\xi_0, \xi_1, \cdots, \xi_k, \eta_0, \eta_1, \cdots, \eta_k,$$

where

$$\xi_j = \alpha_j \phi_j + \beta_j \phi_{r-k+j}, \qquad j = 0, 1, 2, \cdots, k$$

$$\eta_j = (-\beta_j \phi_j + \alpha_j \phi_{r-k+j})e^{\frac{j\pi i}{k}}, \qquad j = 0, 1, 2, \cdots, k$$

with $\alpha_j = \cos\frac{j\pi}{2k}$, $\beta_j = \sin\frac{j\pi}{2k}$.

The new basis is separated into two parts,

$$\cdots \phi_{-2}, \phi_{-1}, \phi_0(= \xi_0), \xi_1, \cdots, \xi_k(= \phi_r), \phi_{r+1}, \cdots$$

and

$$\eta_1, \eta_2, \cdots, \eta_k(= \phi_k), \phi_{k+1}, \cdots, \phi_{r-k}(= \eta_0).$$

Set $B^{(1)}\xi_j = a\xi_j$, $B^{(1)}\eta_j = b\eta_j$, $v^{(1)}\xi_j = \xi_{j+1}$, and $v^{(1)}\eta_j = \eta_{j+1}$. Then under the new basis

$$\{\cdots \phi_{-2}, \phi_{-1}, \phi_0=\xi_0, \xi_1, \cdots, \xi_k=\phi_r, \phi_{r+1}, \cdots\} \bigcup \{\eta_1, \eta_2, \cdots, \eta_k=\phi_k, \phi_{k+1}, \cdots, \phi_{r-k}=\eta_0\},$$

the two matrices

$$
B^{(1)} = \begin{pmatrix} \ddots & & & \\ & a & & \\ & & \ddots & \\ & & & a \\ & & & & \ddots \end{pmatrix} \oplus \begin{pmatrix} b & & \\ & \ddots & \\ & & b \end{pmatrix} = A^{(1)} \oplus C^{(1)}
$$

and

$$
v^{(1)} = \begin{pmatrix} 0 & 0 & \cdots & 1 \\ 1 & 0 & \cdots & 0 \\ \vdots & \ddots & \ddots & \vdots \\ 0 & \cdots & 1 & 0 \end{pmatrix} \oplus \begin{pmatrix} 0 & 0 & \cdots & 1 \\ 1 & 0 & \cdots & 0 \\ \vdots & \ddots & \ddots & \vdots \\ 0 & \cdots & 1 & 0 \end{pmatrix} = u^{(1)} \oplus w^{(1)}
$$

are close to A and u respectively. Actually, $\|A - B^{(1)}\| \le |a - b| = \frac{1}{M}$ and

$$
\|v^{(1)} - u\| \le \sup_j \|(v^{(1)} - u)(\xi_j)\| + \sup_j \|(v^{(1)} - u)(\eta_j)\|.
$$

By the definition of ξ_j,

$$
\begin{aligned}
\|(v^{(1)} - u)(\xi_j)\| &= \|(v^{(1)}(\xi_j) - u(\alpha_j \phi_j + \beta_j \phi_{r-k+j}))\| \\
&= \|\xi_{j+1} - \alpha_j \phi_{j+1} - \beta_j \phi_{r-k+j+1}\| \\
&\le \|\alpha_{j+1} - \alpha_j\| + \|\beta_j - \beta_{j+1}\| \\
&\le \frac{\pi}{2k} + \frac{\pi}{2k} = \frac{\pi}{k} \le \frac{4\pi}{M} \ .
\end{aligned}
$$

(Recall that $\frac{M}{4} \le k \le \frac{M}{2} - 1$.) Similarly, $|(v^{(1)} - u)(\eta_j)| \le \frac{4\pi}{M}$. Therefore,

$$\|v^{(1)} - u\| \le \frac{8\pi}{M}.$$

Now under the new basis, A is decomposed into $B^{(1)} = A^{(1)} \oplus C^{(1)}$ and u is decomposed into $v^{(1)} = u^{(1)} \oplus w^{(1)}$, where $C^{(1)} = bI$ is a constant block and $w^{(1)}$ is a unitary matrix commuting with $C^{(1)}$. $A^{(1)}$ and $u^{(1)}$ have the same properties as what A and u have, respectively. Consider our new pair $(A^{(1)}, u^{(1)})$. We observe that it has one less block at entry b than (A, u). Repeat the above procedure, choose a relative maximal of $A^{(1)}$, and perturb the pair so as to reduce the entry of this block and produce another pair $(A^{(2)}, u^{(2)})$. We observe that the perturbation required at this step is necessary only on the vectors unperturbed at the previous stage, because each perturbed block becomes a central part of a larger block with M unperturbed vectors at each end. Since the only vectors perturbed are those within $\frac{2}{M}$ of a flanking block, we never again perturb the once perturbed block. Therefore, the pair we got after the second step,

$$B^{(2)} = A^{(2)} \oplus C^{(2)} \oplus C^{(1)}, \qquad v^{(2)} = u^{(2)} \oplus w^{(2)} \oplus w^{(1)}$$

satisfy the relations

$$\|A - B^{(2)}\| \le \frac{1}{M} \qquad \text{and} \qquad \|u - v^{(2)}\| \le \frac{8\pi}{M}.$$

We continue in this manner until we are left with a constant matrix

$$A^{(l)} = \begin{pmatrix} \mu_1 & & \\ & \ddots & \\ & & \mu_1 \end{pmatrix}$$

and a unitary matrix

$$u^{(l)} = \begin{pmatrix} 0 & 0 & \cdots & 1 \\ 1 & 0 & \cdots & 0 \\ \vdots & \ddots & \ddots & \vdots \\ 0 & \cdots & 1 & 0 \end{pmatrix},$$

where $A^{(l)}u^{(l)} = u^{(l)}A^{(l)}$. Then the matrices

$$B^{(l)} = A^{(l)} \oplus C^{(l)} \oplus C^{(l-1)} \oplus \cdots \oplus C^{(1)},$$

$$v^{(l)} = u^{(l)} \oplus w^{(l)} \oplus w^{(l-1)} \oplus \cdots \oplus w^{(1)}$$

satisfy

$$\|A - B^{(l)}\| \le \frac{1}{M}, \qquad \|u - v^{(l)}\| \le \frac{8\pi}{M} \qquad \text{and} \qquad B^{(l)}v^{(l)} = v^{(l)}B^{(l)}.$$

Therefore, to prove the theorem, we only need to choose M to satisfy an extra condition, $M > \dfrac{8\pi}{\varepsilon}$.

$\square$

Lemma 3.4. *Given $\varepsilon > 0$ and tree $X \subset \mathbb{C}$, there exists $\delta > 0$ such that, for any diagonal matrix $A \in M_n$ with $SpA \subset X$, and any unitary matrix $u \in M_n$*

which is a permutation satisfying

$$\|uAu^* - A\| < \delta,$$

there is a unitary matrix valued path $u_t \in M_n(C[0,2])$ with the properties $u_0 = I_n$, $u_2 = u$, and

$$\|u_t A u_t^* - A\| < \varepsilon\|A\| + \frac{\varepsilon}{4} \qquad\qquad for\ all\ t \in [0,2].$$

PROOF: Notice that any permutation unitary can be decomposed into a direct sum of the special ones as in Theorem 3.3. By Theorem 3.3, there exists $\delta > 0$ such that if $\|uAu^* - A\| < \delta$, then there exist B and v as in Theorem 3.3 satisfying

$$\|B - A\| < \frac{\varepsilon}{4}, \qquad \|u - v\| < \frac{\varepsilon}{4} \qquad and \qquad vBv^* = B.$$

Let $\delta < \dfrac{\varepsilon}{4}$. Then there exists a continuous unitary valued path u_t, $0 \leq t \leq 1$ such that $u_0 = I_n$, $u_1 = v$ and $u_t B = B u_t$ for all $t \in [0,1]$.

Since $\|u - v\| < \varepsilon/4$, there is a continuous unitary valued path u_t, $1 \leq t \leq 2$ such that $u_1 = v, u_2 = u$ and $\|u_t - u\| < \varepsilon/2$ for all $1 \leq t \leq 2$.

When $0 \leq t \leq 1$,

$$
\begin{aligned}
\|u_t A u_t^* - A\| &= \|u_t(A - B + B)u_t^* - A\| \\
&\leq \|u_t(A - B)u_t^*\| + \|u_t B u_t^* - A\| \\
&= \|A - B\| + \|A - B\| < \frac{\varepsilon}{2},
\end{aligned}
$$

and when $1 \leq t \leq 2$,

$$
\begin{aligned}
\|u_t A u_t^* - A\| &= \|(u_t - u + u)A(u_t - u + u)^* - A\| \\
&\leq \|(u_t - u)A u_t^*\| + \|u A (u_t - u)^*\| + \|u A u^* - A\| \\
&\leq \frac{\varepsilon}{2}\|A\| + \frac{\varepsilon}{2}\|A\| + \delta \\
&\leq \|A\|\varepsilon + \frac{\varepsilon}{4} \ .
\end{aligned}
$$

$\square$

Theorem 3.5. *For any $\varepsilon > 0$ and tree $X \subset \mathbb{C}$, there exists $\delta > 0$ such that the following statement is true: For any graph Y and diagonal matrices*

$$
A(y) = \begin{pmatrix} \alpha_1(y) & & \\ & \ddots & \\ & & \alpha_n(y) \end{pmatrix}, \qquad
B(y) = \begin{pmatrix} \beta_1(y) & & \\ & \ddots & \\ & & \beta_n(y) \end{pmatrix},
$$

where $\alpha_i, \beta_i \in C(Y, X)$, if for each $y_0 \in Y$, there is a permutation $\sigma = \sigma_{y_0} \in S_n$ such that

$$
|\alpha_{\sigma(i)}(y_0) - \beta_i(y_0)| < \delta, \qquad i = 1, 2, \cdots, n,
$$

then there is a unitary $u \in M_n(C(Y))$ satisfying

$$
\|u(y)A(y)u(y)^* - B(y)\| < \varepsilon
$$

for all $y \in Y$.

PROOF: Let $L = \sup\{|z|; \ z \in X \subset \mathbb{C}\} + 1$. First, for $\varepsilon/L > 0$, by Lemma 3.4, choose $\delta_1 > 0$, $\delta_1 < \varepsilon/5$ such that, if $A \in M_n$ is a diagonal matrix with $\mathrm{Sp}(A) \subset X$, and $u \in M_n$ is a permutation unitary with $\|uAu^* - A\| < \delta_1$, then there is a unitary path $u_t \in M_n(C[0,1])$ which connects I_n and u and satisfies

$$\|u_t A u_t^* - A\| < \frac{\varepsilon}{5L}\left(\|A\| + \frac{1}{4}\right) < \frac{\varepsilon}{5}$$

for all $t \in [0,1]$. Choose $\eta > 0$ such that, for any $y_1, y_2 \in Y$, $|y_1 - y_2| < \eta$ implies that

$$|\alpha_i(y_1) - \alpha_i(y_2)| < \frac{\delta_1}{5}, \qquad\qquad |\beta_i(y_1) - \beta_i(y_2)| < \frac{\delta_1}{5},$$

for all $i = 1, 2, \cdots, n$. Let $T \subset Y$ be an $\frac{\eta}{2}$-dense finite subset which contains all the joint points of Y. Then Y is divided into small intervals by the points in T, the length of each small interval being shorter than $\frac{\eta}{2}$. Let Γ denote the set of all such intervals.

Set $\delta = \frac{\delta_1}{5}$. By the assumption, for each $t_k \in T$, there is a permutation $\sigma_{t_k} \in S_n$ such that

$$|\alpha_{\sigma_{t_k}(i)}(t_k) - \beta_i(t_k)| < \delta = \frac{\delta_1}{5}, \qquad\qquad i = 1, 2, \cdots, n.$$

Define u_{t_k} to be the permutation unitary corresponding to σ_{t_k}, i.e.,

$$u_{t_k} \begin{pmatrix} \alpha_1(t_k) & & \\ & \ddots & \\ & & \alpha_n(t_k) \end{pmatrix} u_{t_k}^* = \begin{pmatrix} \alpha_{\sigma_{t_k}(1)}(t_k) & & \\ & \ddots & \\ & & \alpha_{\sigma_{t_k}(n)}(t_k) \end{pmatrix}.$$

We now have the definition of u_t on the set $T \subset Y$. In order to define u_t on the

whole set Y, take any small interval $\gamma \in \Gamma, \gamma = [t_1, t_2] \subset Y$. We have

$$\|u_{t_1} A(t_1) u_{t_1}^* - B(t_1)\| < \delta = \frac{\delta_1}{5}, \qquad \|u_{t_2} A(t_2) u_{t_2}^* - B(t_2)\| < \frac{\delta_1}{5}.$$

Then

$$\|u_{t_2}^* u_{t_1} A(t_1) u_{t_1}^* u_{t_2} - A(t_1)\|$$

$$= \|u_{t_1} A(t_1) u_{t_1}^* - u_{t_2} A(t_1) u_{t_2}^*\|$$

$$\leq \|u_{t_1} A(t_1) u_{t_1}^* - u_{t_2} A(t_2) u_{t_2}^*\| + \|A(t_2) - A(t_1)\|$$

$$\leq \|u_{t_1} A(t_1) u_{t_1}^* - B(t_1)\| + \|u_{t_2} A(t_2) u_{t_2}^* - B(t_2)\| +$$

$$+ \|B(t_1) - B(t_2)\| + \|A(t_1) - A(t_2)\|$$

$$< \frac{\delta_1}{5} + \frac{\delta_1}{5} + \frac{\delta_1}{5} + \frac{\delta_1}{5} < \delta_1.$$

By Lemma 3.4, there is a unitary valued path $v_t \in M_n(C[t_1, t_2])$ such that $v_{t_1} = u_{t_2}^* u_{t_1}$, $v_{t_2} = I_n$ and

$$\|v_t A(t_1) v_t^* - A(t_1)\| < \frac{\varepsilon}{5} \qquad\qquad \text{for all } t \in [t_1, t_2].$$

Let $u_t = u_{t_2} v_t, t \in [t_1, t_2]$. Then $u_t = u_{t_1}$ when $t = t_1$ and $u_t = u_{t_2}$ when $t = t_2$.

And for $t \in [t_1, t_2]$,

$$\|u_t A(t) u_t^* - B(t)\|$$

$$\leq \quad \|u_t(A(t) - A(t_1))u_t^*\| + \|u_t A(t_1) u_t^* - u_{t_2} A(t_1) u_{t_2}^*\| +$$

$$+\|u_{t_2}(A(t_1) - A(t_2))u_{t_2}^*\| + \|u_{t_2} A(t_2) u_{t_2}^* - B(t_2)\| + \|B(t_2) - B(t)\|$$

$$< \quad \varepsilon.$$

$\square$

Combining Corollary 2.1.7 and Theorem 3.5, we have the following corollary:

Corollary 3.6. *For any $\varepsilon > 0$ and any tree $X \subset \mathbb{C}$, there exists $\delta > 0$ such that, if unital homomorphisms $\phi, \psi : C(X) \to M_n(C(Y))$ (Y is also a tree in $\mathbb{C}$) satisfy the condition that, for each $y \in Y$, the two sets $Sp\phi_y$ and $Sp\psi_y$ can be paired within δ, then there is a unitary $u \in M_n(C(Y))$ satisfying*

$$\|\phi(h) - Adu \circ \psi(h)\| < \varepsilon,$$

where h is the generator of $C(X)$ with $h(x) = x$.

Chapter 4

Approximate Divisibility

In this chapter, we will prove the approximate divisibility of a simple inductive limit C^*-algebra of finite direct sums of matrix algebras over trees. This property will be used in the proof of the Existence Theorem.

Definition 4.1. (See [BKR].) A C^*-algebra A is said to be **approximately divisible** if, for any finite subset $F \subset A$, any $\varepsilon > 0$, and any integer $N > 0$, there is a finite dimensional unital sub-C^*-algebra $A_0 \subset A$ and a finite subset $F_0 \subset A$, F_0 commuting with A_0, such that, for any $f \in F$, $\mathrm{dist}(f, F_0) < \varepsilon$, i.e., $F \subset_\varepsilon F_0$, and such that each simple direct summand of A_0 is of order at least N.

4.2. Suppose that C^*-algebra A is an inductive limit of a sequence

$$A_1 \xrightarrow{\phi_{1,2}} A_2 \xrightarrow{\phi_{2,3}} A_3 \longrightarrow \cdots,$$

where the algebras $A_n = \oplus_{i=1}^{k_n} M_{[n,i]}(C(X_{n,i}))$, and the spaces $X_{n,i}$ are trees. The following property of the inductive limit system can be regarded as a local version of the approximate divisibility which implies that A is approximately divisible. For any $\varepsilon > 0$, two integers $n > 0$ and $N > 0$, and any finite subset $F \subset A_n$, there exists $M > n$ such that, for any $m \geq M$, there exists a map

$$\psi: \quad A_n \longrightarrow A_m$$

with the following properties: For any i, j, there exist continuous functions $\{\lambda_s(y)\}_{s=1}^p$ from $X_{m,j}$ to $X_{n,i}$, a unitary $u \in A_m^j$, and integers $l_s \geq N$, $s = 1, 2, \cdots, p$ such that the partial map

$$\psi^{i,j}: \quad A_n^i \ (= M_{n,i}(C(X_{n,i}))) \longrightarrow A_m^j$$

can be expressed as

$$\psi^{i,j}(f)(y) = u \begin{pmatrix} 0 & & & & \\ & f(\lambda_1(y)) \otimes I_{l_1} & & & \\ & & \ddots & & \\ & & & f(\lambda_p(y)) \otimes I_{l_p} & \\ & & & & 0 \end{pmatrix} u^*$$

for all $y \in X_{m,j}$, where, for each $i = 1, 2, \cdots, p$,

$$f(\lambda_i(y)) \otimes I_{l_i} := \mathrm{diag}(\underbrace{f(\lambda_i(y)), \cdots, f(\lambda_i(y))}_{l_i-copies}) = \begin{pmatrix} f(\lambda_i(y)) & & \\ & \ddots & \\ & & f(\lambda_i(y)) \end{pmatrix}.$$

Furthermore, ψ satisfies

$$\|\phi_{n,m}(f) - \psi(f)\| < \varepsilon$$

for all $f \in F$. We will prove the above property for the case that the limit algebra A is simple ($A \neq M_n(\mathbb{C})$) and the spaces $X_{n,i}$ are trees.

Theorem 4.3. *Given $\varepsilon > 0$, a tree $X \subseteq \mathbb{C}$ with $X \neq \{pt\}$, an integer $N > 0$, and a finite set $F \subset C(X)$, there exists $\delta > 0$ such that, for any unital homomorphism $\phi: C(X) \longrightarrow M_q(C(Y))$, where Y is a tree, if $Sp\phi_y$ is δ-dense in X for each $y \in Y$, then there is a unital homomorphism $\psi: C(X) \to M_q(C(Y))$ satisfying the following conditions:*

(i) $\|\psi(f) - \phi(f)\| < \varepsilon$ for all $f \in F$;

(ii) There are continuous functions $\lambda_1, \lambda_2, \cdots, \lambda_p \in C(Y, X)$ and a unitary $u \in M_q(C(Y))$ such that, for every $y \in Y$,

$$\psi(f)(y) = u(y) \begin{pmatrix} f(\lambda_1(y)) \otimes I_{l_1} & & \\ & \ddots & \\ & & f(\lambda_p(y)) \otimes I_{l_p} \end{pmatrix} u(y)^*$$

with $l_s \geq N$, $1 \leq s \leq p$.

4.4. The proof will be divided into several steps (from 4.4 to 4.8). By Theorem 2.1.7, with a small perturbation, ϕ can be changed to ϕ_1 which is of diagonal form. That is, there are continuous functions $x_1, \cdots, x_q \in C(Y, X)$ and a unitary $v \in M_q(C(Y))$ such that

$$
\phi_1(f)(y) = v(y) \begin{pmatrix} f(x_1(y)) & & \\ & \ddots & \\ & & f(x_q(y)) \end{pmatrix} v(y)^*
$$

for all $y \in Y$ and $f \in C(X)$. Furthermore, $x_1, \cdots, x_q$ are distinct. So we may simply assume that ϕ itself is of this form. We assume that, for each $y \in Y$, $\mathrm{Sp}\phi_y = \{x_1(y), \cdots, x_q(y)\}$ is δ-dense in X (where δ is to be determined later).

For any $\zeta > 0$ (ζ is also to be determined later), there is a $\eta > 0$ such that, for any $y_1, y_2 \in Y$, $\mathrm{dist}(y_1, y_2) < \eta$ implies that

$$
\mathrm{dist}(x_i(y_1), x_i(y_2)) < \frac{\zeta}{10}, \qquad\qquad i = 1, 2, \cdots, q
$$

for all $x_i \in \mathrm{Sp}\phi_y = \{x_1(y), \cdots, x_q(y)\}$.

Let m be a positive integer with $\frac{1}{m} < \eta$. Identify every edge of X and Y as the interval $[0, 1]$. We will split each edge of Y, say, $[0, 1] \subset Y$, into m sub-intervals of equal length, $[0, \frac{1}{m}]$, $[\frac{1}{m}, \frac{2}{m}]$, $[\frac{m-1}{m}, 1]$. Evaluating at each splitting point $\frac{i}{m}$, let us consider the set $\{x_1(\frac{i}{m}), \cdots, x_q(\frac{i}{m})\}$. This set is δ-dense in X. First, for each fixed i, we wish to group the points this set into $p + 1$ groups, where $q = (p + 1)N + r$. Each group has N points, except one, which has $N + r$ points, such that the diameter of each group is at most $2N\delta$.

Fix an extreme point $P_0 \in X$ as a starting point. Let us consider the set $\{x_1(0), \cdots, x_q(0)\} \subset X$. Write it as $\bigcup_{i=0}^{p} F_i$, as follows. F_0 is the set consisting of the first $N + r$ points in X starting from the point $P_0 \in X$. Each set F_i contains exactly N points in $\{x_1(0), \cdots, x_q(0)\}$. The diameter of each F_i is less than $2N\delta$. This is possible since we may choose the first $N + r$ points of $x_i(0)$'s starting from $P_0 \in X$, the special vertex of X fixed above, as F_0, then the next N points in this edge, as our F_1, and so on, until we meet the joint point Q.

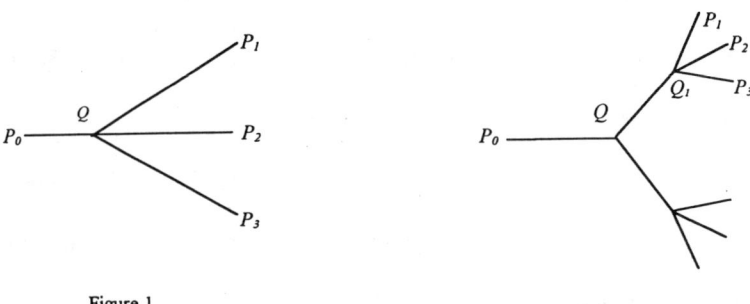

Figure 1   Figure 2

Suppose that we do not have enough N points left before the joint point Q, i.e., we have $r_0 < N$ points left. Let X be the tree in figure 1, i.e., $X \setminus [P_0, Q]$ has three components $(Q, P_1]$, $(Q, P_2]$, $(Q, P_3]$. Then on each edge $(Q, P_i]$ we may group the points similarly as on $[P_0, Q]$, group the first N points in $(Q, P_i]$, starting from P_i, then the next N points, and so on, until we have no enough N points left before the joint point Q. Suppose that there are r_i points left, $i = 1, 2, 3$. Then $r_0 + r_1 + r_2 + r_3 = SN$ for some integer S. We group these SN points as S sets, each of which being with N points. Then we get

$$\{x_1(0), x_2(0), \cdots, x_q(0)\} = \bigcup_{i=0}^{p} F_i.$$

If X is more complicated like the one in figure 2, then we may group $\mathrm{Sp}\phi_0 = \{x_1(0), \cdots, x_q(0)\}$ by starting from all the other extreme points of X, $P_1, P_2, P_3, \cdots$, until we meet at Q_1 (other extreme points may meet at different joint points which should be dealt with separately but similarly). If r_1, r_2, r_3 are the numbers of the points left on $(Q_1, P_1]$, $(Q_1, P_2]$, $(Q_1, P_3]$, respectively, $(r_i < N)$, then there is an $r_4 < N$, $r_4 \geq 0$ such that $r_1 + r_2 + r_3 + r_4 = SN$ for some $S > 0$, (i.e., a multiple of N). Take r_4 points on $(Q, Q_1]$, starting from Q_1, and group all these $r_1 + r_2 + r_3 + r_4$ points as S sets, each with N points. Then group the remaining points on $(Q, Q_1]$, one by one, from Q_1 to Q. It is easy to see that, by this procedure, we can group $\mathrm{Sp}\phi_0$ as $\bigcup_{i=0}^{p} F_i$, with $|F_0| = N + r$, and $|F_i| = N$ for all $i \geq 1$. (Notice that we can give a partial order to the tree X if we fix a starting point P_0. Since each vertex other than P_0 has exactly one previous vertex point, the tree in figure 2 has the general feature of arbitrary trees.)

For any $x, y \in F_0$, $\mathrm{dist}(x, y) \leq (N + r)\delta \leq 2N\delta$. And for any $x, y \in F_i$, with $i \geq 1$, there are two cases. One is that all the points in F_i are contained in one edge of X. In this case, $\mathrm{dist}(x, y) \leq N\delta$. Another one is that x and y belong to two different edges of X with a common joint point J. Then

$$\mathrm{dist}(x, y) \leq \mathrm{dist}(x, J) + \mathrm{dist}(J, y) \leq 2N\delta.$$

Hence, in all the cases, $\mathrm{dist}(x, y) \leq 2N\delta$ for any $x, y \in F_i$, $i = 0, 1, \cdots, p$.

Make the above grouping procedure on each set $\{x_1(\frac{i}{m}), x_2(\frac{i}{m}), \cdots, x_q(\frac{i}{m})\}$ for all splitting points $\{\frac{i}{m}\}_{i=0}^{m} \subset [0, 1]$ and for all edges of Y.

For each splitting point $\frac{k}{m} \in [0,1]$ and the grouping at this point, denoted by

$$F_0(\frac{k}{m}) \bigcup F_1(\frac{k}{m}) \bigcup \cdots \bigcup F_p(\frac{k}{m}) \, ,$$

we will choose a point $a_i \in F_i(\frac{k}{m})$ as a representative, for each $i = 0, 1, \cdots, p$. This gives us a subset $\{a_i\}_0^p \subset \mathrm{Sp}\phi_{\frac{k}{m}}$. If $\{b_i\}_0^p$ is the set chosen in this way at the point $\frac{k+1}{m}$, we will prove that $\{a_i\}_0^p$ and $\{b_i\}_0^p$ can be paired within $c = 4N\delta + \frac{\varsigma}{10}$ for all k.

Without loss of generality, we consider only the case $k = 0$. That is, we will prove that the two sets $\{a_i\}_0^p \subset \mathrm{Sp}\phi_0$ and $\{b_i\}_0^p \subset \mathrm{Sp}\phi_{\frac{1}{m}}$ can be paired within $c = 4N\delta + \frac{\varsigma}{10}$.

Lemma 4.5. *Suppose that our groupings at points* $0, \frac{1}{m} \in I_\alpha$ *are*

$$Sp\phi_0 = \{x_1(0), x_2(0), \cdots, x_q(0)\} = \bigcup_{i=0}^p F_i \, ,$$

$$Sp\phi_{\frac{1}{m}} = \{x_1(\tfrac{1}{m}), x_2(\tfrac{1}{m}), \cdots, x_q(\tfrac{1}{m})\} = \bigcup_{i=0}^p G_i \, ,$$

respectively. Choose an arbitrary $a_i \in F_i$ *and an arbitrary* $b_i \in G_i$ *as representatives for each* $i = 0, 1, \cdots, p$. *Then there is a bijection*

$$\tau : \quad \{a_i\}_0^p \longrightarrow \{b_i\}_0^p$$

such that $\tau(a_0) = b_0$ *and* $dist(a, \tau(a)) < c = 4N\delta + \frac{\varsigma}{10}$ *for all* $a \in \{a_i\}_0^p$.

PROOF: Consider the groupings $\bigcup_{i=0}^p F_i = \{x_1(0), x_2(0), \cdots, x_q(0)\}$, and $\bigcup_{i=0}^p G_i = \{x_1(\tfrac{1}{m}), x_2(\tfrac{1}{m}), \cdots, x_q(\tfrac{1}{m})\}$. We claim that if $F_0 = \{x_{i_1}(0), x_{i_2}(0), \cdots, x_{i_{N+r}}(0)\}$, then $G_0 \subset \{x_1(\tfrac{1}{m}), x_2(\tfrac{1}{m}), \cdots, x_q(\tfrac{1}{m})\}$ must be equal to $\{x_{i_1}(\tfrac{1}{m}), x_{i_2}(\tfrac{1}{m}), \cdots, x_{i_{N+r}}(\tfrac{1}{m})\}$.

First, notice that when we make our groupings at each point $\frac{i}{m}$, we always start from one fixed extreme point P_0 of X. Hence our first set F_0 (or G_0) is the only one with $N + r$ points. (We can assume that, at each edge of X, there are at least $2N$ points of $\mathrm{Sp}\phi_y$ for each $y \in Y$ because, for each $y \in Y$, $\mathrm{Sp}\phi_y$ is supposed to be δ-dense in X, and $\delta < \frac{1}{2N}$.) Therefore, at each $y = \frac{i}{m}$, if our grouping is $\bigcup_{i=0}^{p} H_i = \{x_1(\frac{i}{m}), x_2(\frac{i}{m}), \cdots, x_q(\frac{i}{m})\}$, then the first group has the following expression:

$$H_0 = \{x_{i_1}(\frac{j}{m}), x_{i_2}(\frac{j}{m}), \cdots, x_{i_{N+1}}(\frac{j}{m})\} \ .$$

(Notice that the distinctness of $\{x_1(y), \cdots, x_p(y)\}$ for all $y \in Y$ implies that, if $x_{i_1}(0)$ is the first point starting with the specified extreme point P_0, then accordingly, $x_{i_1}(y)$ is the first point starting with P_0 for all $y \in Y$.)

Now in each i, choose an element $a_i \in F_i$ and an element $b_i \in G_i$. From the above argument, we may require that there exists an eigenvalue $x_{i_1} \in \mathrm{Sp}\phi$ with $x_{i_1}(0) = a_0$, $x_{i_1}(\frac{1}{m}) = b_0$. Set $A = \{a_1, a_2, \cdots, a_p\}$ and $B = \{b_1, b_2, \cdots, b_p\}$. We will prove that there exists a pairing $\alpha : A \to B$ satisfying

$$\mathrm{dist}(\alpha(a), \ a) \ < \ c = 4N\delta + \frac{\zeta}{10} \qquad\qquad \text{for all } \ a \in A \ .$$

First, $|A| = |B| = p$. Recall that we use $|E|$ to denote the cardinal number of the elements of a set E. For each $a \in A$, set $B(a) = \{b_i \in B; \ \mathrm{dist}(a, \ b_i) < c\} = B_c(a) \cap B$. And for a subset $A_1 \subset A$, set $B(A_1) = \bigcup_{a \in A_1} B(a)$. We will prove that $|B(A_1)| \geq |A_1|$ for each $A_1 \subset A$. Let $A_1 = \{a_{i_1}, a_{i_2}, \cdots, a_{i_t}\} \subset A$, $a_{i_k} \in F_k$. Note that $|F_{i_1} \bigcup F_{i_2} \bigcup \cdots \bigcup F_{i_t}| = t \cdot N$. (Each F_{i_k} is of N elements.) Reordering the set

A we may assume that

$$F_{i_1} \bigcup F_{i_2} \bigcup \cdots \bigcup F_{i_t} = \{x_1(0), x_2(0), \cdots, x_{tN}(0)\} \subset \text{Sp}\phi_0 .$$

Then $\{x_1(\frac{1}{m}), x_2(\frac{1}{m}), \cdots, x_{tN}(\frac{1}{m})\} \bigcap G_0 = \varnothing$. Let

$$\{G_{j_1}, G_{j_2}, \cdots, G_{j_s}\} = \{G_j; \quad G_j \cap \{x_1(\frac{1}{m}), x_2(\frac{1}{m}), \cdots, x_{tN}(\frac{1}{m})\} \neq \varnothing\}.$$

Then it is evident that $s \geq t$. From $G_j \cap \{x_1(\frac{1}{m}), x_2(\frac{1}{m}), \cdots, x_{tN}(\frac{1}{m})\} \neq \varnothing$ we know

that $G_j \subset B_{2N\delta}(\{x_1(\frac{1}{m}), x_2(\frac{1}{m}), \cdots, x_{tN}(\frac{1}{m})\})$, since the diameter of G_j is less than

$2N\delta$ for all j. We have

$$\begin{aligned}
\{b_{j_1}, b_{j_2}, \cdots, b_{j_s}\} \quad &\subset \quad G_{j_1} \bigcup G_{j_2} \bigcup \cdots \bigcup G_{j_s} \\
&\subset \quad B_{2N\delta}(\{x_1(\frac{1}{m}), x_2(\frac{1}{m}), \cdots, x_{tN}(\frac{1}{m})\}) \\
&\subset \quad B_{2N\delta + \frac{c}{10}}(\{x_1(0), x_2(0), \cdots, x_{tN}(0)\}) \\
&\subset \quad B_{2N\delta + \frac{c}{10} + 2N\delta}(\{a_{i_1}, a_{i_2}, \cdots, a_{i_t}\}) = B_c(A_1) .
\end{aligned}$$

Hence for any $A_1 \subset A$,

$$|B(A_1)| = s \geq t = |A_1| .$$

Notice that $A_1 \subset A$ is arbitrary, and the positions of the sets A and B are symmetric in the above argument. By Marriage Lemma in [HV], A and B can be paired within $4N\delta + \frac{c}{10}$. That is, there is a one to one map $\tau : \quad A \to B$ such that $\text{dist}(a, \tau(a)) < c$ for all $a \in A$.

$\square$

4.6. Set $H_i^j = F_i(\frac{j}{m}) \subset \mathrm{Sp}\phi_{\frac{j}{m}}$ for $\frac{j}{m} \in [0,1] \subset Y$, with H_0^j the first set containing $N + r$ points. Let $a_i^j \in H_i^j$ be the fixed point in H_i^j. By Lemma 4.5, there exists a bijection

$$\tau_j : \ \{a_1^{j-1}, a_2^{j-1}, \cdots, a_p^{j-1}\} \longrightarrow \{a_1^j, a_2^j, \cdots, a_p^j\}$$

for each $j = 1, 2, \cdots, m$ and for all edges $[0,1] \subset Y$. Our next step is to connect each pair $\{a_i^{j-1}, \tau(a_i^j)\}$ by a path in X so that we may construct p continuous functions from Y to X with certain conditions we desired. Here we need another lemma below:

Lemma 4.7. Let $\lambda_i : \ [0, \frac{1}{m}] \subset Y \to X$ denote the shortest path in X, connecting a_i and $\tau(a_i) \in \{b_i\}_{i=0}^p$. And let

$$E_i(y) = \ \underbrace{\{\lambda_i(y), \lambda_i(y), \cdots, \lambda_i(y)\}}_{l_i-copies}, \qquad y \in [0, \frac{1}{m}], \ i = 0, 1, \cdots p \ ,$$

where $l_i = N$ when $i \geq 1$ and $l_0 = N + r$. Then, for each $y \in [0, \frac{1}{m}]$, the two sets

$$\mathrm{Sp}\phi_y = \{x_1(y), x_2(y), \cdots, x_q(y)\}$$

and

$$E_0(y) \bigcup E_1(y) \bigcup \cdots \bigcup E_p(y)$$
$$= \ \{ \underbrace{\lambda_0(y), \cdots, \lambda_0(y)}_{N+r}, \ \underbrace{\lambda_1(y), \cdots, \lambda_1(y)}_{N}, \ \cdots, \ \underbrace{\lambda_p(y), \cdots, \lambda_p(y)}_{N} \ \}$$

can be paired within $C_1 := 6N\delta + \dfrac{\zeta}{5}$.

PROOF: Set $F_i(y) = \{x_j(y) \in \mathrm{Sp}\phi_y;\quad x_j(0) \in F_i\}$. Here we still use the notation in Lemma 4.5. Then

$$F_0(y)\bigcup F_1(y)\bigcup\cdots\bigcup F_p(y) = \mathrm{Sp}\phi_y$$

for all $y \in [0, \frac{1}{m}]$. And set $F_i(0) = F_i$ as in Lemma 4.5. We need to prove that the sets $F_0(y)\bigcup F_1(y)\bigcup\cdots\bigcup F_p(y)$ and $E_0(y)\bigcup E_1(y)\bigcup\cdots\bigcup E_p(y)$ can be paired within $C_1 = 6N\delta + \frac{\zeta}{5}$ for all $y \in [0, \frac{1}{m}] \subset Y$.

Notice that $|F_i(y)| = |F_i| = |E_i(y)|$, $i = 0, 1, \cdots, p$. For any $x_k(y) \in F_i(y)$, we have $x_k(0) \in F_i$. Hence $|x_k(0) - a_i| < 2N\delta$, where $a_i \in F_i$ was chosen in Lemma 4.5. So

$$|x_k(y) - a_i| \le |x_k(y) - x_k(0)| + |x_k(0) - a_i| < \frac{\zeta}{10} + 2N\delta.$$

For $\lambda_i(y) \in E_i(y)$, we have

$$
\begin{aligned}
\mathrm{dist}(x_k(y), \lambda_i(y)) &\le \mathrm{dist}(x_k(y), a_i) + \mathrm{dist}(a_i, \lambda_i(y)) \\
&\le \frac{\zeta}{10} + 2N\delta + \mathrm{dist}(a_i, \tau(a_i)) \\
&\le \frac{\zeta}{10} + 2N\delta + 4N\delta + \frac{\zeta}{10} \qquad \text{(Lemma 4.5)} \\
&\le 6N\delta + \frac{\zeta}{5} = C_1.
\end{aligned}
$$

$\square$

4.8. Our aim is to construct q functions $\mu_1, \mu_2, \cdots, \mu_q : Y \to X$ with the following properties:

(i) For each $y \in Y$, the sets $\mathrm{Sp}\phi_y = \{x_1(y), x_2(y), \cdots, x_q(y)\}$ and

$\{\mu_1(y), \mu_2(y), \cdots, \mu_q(y)\}$ can be paired within $6N\delta + \dfrac{\zeta}{5}$;

(ii) For all $y \in Y$,

$$\mu_1(y) = \mu_2(y) = \cdots = \mu_{N+r}(y)$$

$$\mu_{N+r+1}(y) = \mu_{N+r+2}(y) = \cdots = \mu_{2N+r}(y)$$

$$\vdots$$

$$\mu_{q-N+1}(y) = \mu_{q-N+2}(y) = \cdots = \mu_q(y).$$

For this purpose, let us give the notation globally. Let $Y = \bigcup_{\alpha=1}^\Lambda I_\alpha$, where each $I_\alpha = [0,1]$ is an edge of Y. For each fixed α and $k \le m$, let $\{G_i^{\alpha,k}\}_{i=0}^p$ denote the grouping of $\mathrm{Sp}\phi_y$ for $y = \frac{k}{m} \in I_\alpha$, assuming $G_0^{\alpha,k}$ to be the first $N + r$ points in $\{x_1(\frac{k}{m}), \cdots, x_q(\frac{k}{m})\} \subset X$ starting from the special point $P_0 \in X$. Fix $a_i^{\alpha,k} \in G_i^{\alpha,k}$ for all i, k, α. For convenience, we choose $a_0^{\alpha,k}$ to be the point in $\{x_1(\frac{k}{m}), \cdots, x_q(\frac{k}{m})\} \subset X$ which is the closest one to the point $P_0 \in X$. By Lemma 4.5, for each α and k, there exists a bijection

$$\tau^{\alpha,k} : \{a_i^{\alpha,k}\}_{i=0}^p \longrightarrow \{a_i^{\alpha,k+1}\}_{i=0}^p$$

such that $\tau^{\alpha,k}(a_0^{\alpha,k}) = a_0^{\alpha,k+1}$ and

$$\mathrm{dist}(\tau^{\alpha,k}(a_i^{\alpha,k}), a_i^{\alpha,k}) < c = 4N\delta + \frac{\zeta}{10}$$

for $i = 0, 1, \cdots, p$. Let us start from an extreme point Q_0 of Y. Let $Q_0 \in I_{\alpha_0} = [0,1] \subset Y$. Define $\lambda_i(Q_0) = a_i^{\alpha_0,0}$, $i = 0, 1, \cdots, p$. and define $\lambda_i(\frac{1}{m}) = \tau^{\alpha_0,0}(a_i^{\alpha_0,0})$, $i = 0, 1, \cdots, p$. By this method, it is easily seen that once λ_i is de-

fined at point $\frac{k}{m} \in [0,1] = I_\alpha$, it is defined at $\frac{k+1}{m} \in [0,1] = I_\alpha$ $(k+1 \le m)$.
Thus we have the definition of λ_i on all the splitting points $\frac{k}{m} \in [0,1] = I_\alpha \subset Y$,
$i = 0, 1, \cdots, p$.

Now define λ_i on $[\frac{k}{m}, \frac{k+1}{m}] \subset I_\alpha$ to be the shortest path inside X connecting
$\lambda_i(\frac{k}{m})$ and $\lambda_i(\frac{k+1}{m})$ for all k, α, and $i = 0, 1, \cdots, p$. Then we get the definition of
λ_i on the whole Y, for each $i = 0, 1, \cdots, p$. Define

$$\mu_1(y) = \mu_2(y) = \cdots = \mu_{N+r}(y) \qquad = \lambda_0$$

$$\mu_{N+r+1}(y) = \mu_{N+r+2}(y) = \cdots = \mu_{2N+r}(y) \quad = \lambda_1$$

$$\vdots$$

$$\mu_{q-N+1}(y) = \mu_{q-N+2}(y) = \cdots = \mu_q(y) \qquad = \lambda_p.$$

By Lemma 4.7, the sets $\mathrm{Sp}\phi_y = \{x_1(y), x_2(y), \cdots, x_q(y)\}$ and $\{\mu_1(y), \mu_2(y), \cdots, \mu_q(y)\}$ can be paired within $6N\delta + \frac{\zeta}{5}$. Let $u \in M_q(C(Y))$ be a unitary satisfying

$$\phi(f)(y) = u(y) \begin{pmatrix} f(x_1(y)) & & \\ & \ddots & \\ & & f(x_q(y)) \end{pmatrix} u(y)^*.$$

for all $f \in C(X)$. For convenience, set

$$\alpha(y) = \begin{pmatrix} x_1(y) & & & \\ & x_2(y) & & \\ & & \ddots & \\ & & & x_q(y) \end{pmatrix} \quad \text{and} \quad \beta(y) = \begin{pmatrix} \mu_1(y) & & & \\ & \mu_2(y) & & \\ & & \ddots & \\ & & & \mu_q(y) \end{pmatrix}.$$

Then α, $\beta \in M_q(C(Y))$ are normal elements, and $\phi(f) = \mathrm{Ad}u(f(\alpha))$ for all $f \in C(X)$.

For the finite set $F = \{f_1, f_2, \cdots, f_s\} \subset C(X)$ and $\varepsilon > 0$, given in the theorem, there is an $\varepsilon_1 > 0$ such that, for two diagonal matrices $\alpha_1, \alpha_2 \in M_q(C(Y))$ with spectra contained in X and a unitary $w \in M_q$, if $\|\alpha_1 - \mathrm{Ad}w(\alpha_2)\| < \varepsilon_1$, then

$$\|f(\alpha_1) - f(\mathrm{Ad}w(\alpha_2))\| < \varepsilon$$

for all $f \in F$. (Here $f(\alpha_i)$ is well defined by the theory of functional calculus.) In fact, this is true for f being polynomials. On the other hand, all $f \in C(X)$ can be approximated by polynomials. In addition, $\|f(\alpha) - p(\alpha)\| \leq \|f - p\| \cdot \|\alpha\|$ for all $\alpha \in M_q(\mathbb{C})$ with the properties described above. Perturbing F by a very small number we may assume that all $f_i \in F$ are polynomials.

By Theorem 3.5, there is an $\varepsilon_2 > 0$ such that, if the sets $\{x_1(y), x_2(y), \cdots, x_q(y)\}$ and $\{\mu_1(y), \mu_2(y), \cdots, \mu_q(y)\}$ can be paired within ε_2 for every $y \in Y$, then there s a unitary $v \in M_q(C(Y))$ satisfying

$$\|\alpha(y) - \mathrm{Ad}v(y)(\beta(y))\| < \varepsilon_1$$

or all $y \in Y$. Hence

$$\|f(\alpha(y)) - \mathrm{Ad}v(y)(f(\beta(y)))\| < \varepsilon \qquad (*)$$

for all $y \in Y$ and $f \in F$. Define $\psi(f) = \mathrm{Ad}(uv)(f(\beta))$ for $f \in C(X)$. Then

$$\psi(f)(y) = u(y)v(y) \begin{pmatrix} f(\mu_1(y)) & & & \\ & f(\mu_2(y)) & & \\ & & \ddots & \\ & & & f(\mu_q(y)) \end{pmatrix} v(y)^* u(y)^*$$

$$= u(y)v(y) \begin{pmatrix} f(\lambda_1(y))I_{N+r} & & & \\ & f(\lambda_2(y))I_N & & \\ & & \ddots & \\ & & & f(\lambda_p(y))I_N \end{pmatrix} v(y)^* u(y)^*.$$

And $(*)$ means that

$$\|\phi(f) - \psi(f)\| < \varepsilon$$

for all $f \in F$. Therefore, if we choose our δ and ζ to be small enough such that $6N\delta + \dfrac{\zeta}{5} < \varepsilon_2$, then

$$\|\phi(f) - \psi(f)\| < \varepsilon$$

for all $f \in F$. The proof of Theorem 4.3 is completed.

Remark 4.9. If the tree $X = \{pt\}$ (i.e., a single point set), then Theorem 4.3 is not true. But in this case, if $q \geq N$, then ϕ itself has the form of ψ.

Now we are considering the general case. Let $\phi \colon A \to B$ be a homomorphism, where

$$A = M_{n_1}(C(X_1)) \oplus \cdots \oplus M_{n_s}(C(X_s)) \,,$$

$$B = M_{m_1}(C(Y_1)) \oplus \cdots \oplus M_{m_t}(C(Y_t)) \,,$$

and each of X_i and Y_j is a connected tree (it may be the set of a single point $\{pt\}$). Passing to quotients, we can always assume the second algebra B to be a single block. The following is the theorem in general form.

Theorem 4.10. *Let* $A = M_{n_1}(C(X_1)) \oplus \cdots \oplus M_{n_s}(C(X_s))$. *For any* $\varepsilon > 0$, *any finite set* $F \subset A$, *and any integer* $N > 0$, *there exists* $\delta > 0$ *such that the following statement holds.*

Let $\phi: A = A_1 \oplus A_2 \oplus \cdots \oplus A_s \to B = M_m(C(Y))$ *be a unital homomorphism. Suppose that* $Sp\phi_y$ *is* δ-*dense in* X *for all* $y \in Y$, *and that* $\dfrac{rank\ \phi(1_{A_i})}{rank(1_{A_i})} \geq N$ *for all* $i = 1, 2, \cdots, s$. *Then there exists a unital homomorphism* $\psi : A \to B$ *satisfying the following conditions:*

(i) $\|\psi(f) - \phi(f)\| < \varepsilon$ *for all* $f \in F$;

(ii) For each $i = 1, 2, \cdots, s$, $\psi_i = \psi|_{A_i}$ is of the form

$$
\psi_i(f)(y) \;=\; u(y)
\begin{pmatrix}
0 & & & & & \\
& f(\lambda_1(y)) \otimes I_{l_1} & & & & \\
& & f(\lambda_2(y)) \otimes I_{l_2} & & & \\
& & & \ddots & & \\
& & & & f(\lambda_p(y)) \otimes I_{l_p} & \\
& & & & & 0
\end{pmatrix}
u(y)^*,
$$

where $l_j \geq N$; $1 \leq j \leq p$ and $u \in B$ is a unitary.

PROOF: Without loss of generality, we assume $s = 2$. Consider the units $\mathbf{1}_{A_1} \in A_1$ and $\mathbf{1}_{A_2} \in A_2$. $\phi(\mathbf{1}_{A_1})$ and $\phi(\mathbf{1}_{A_2})$ are orthogonal projections and $\phi(\mathbf{1}_{A_1}) + \phi(\mathbf{1}_{A_2}) = \mathbf{1}_B$. Since $\dim(Y) = 1$, all the projections in $M_q(C(Y))$ are trivial. So there is a unitary $u \in B$ such that

$$
(\mathrm{Ad}u \circ \phi)(\mathbf{1}_{A_1}) =
\begin{pmatrix}
I & 0 \\
0 & 0
\end{pmatrix}
:= E_1 \quad,
$$

$$
(\mathrm{Ad}u \circ \phi)(\mathbf{1}_{A_2}) =
\begin{pmatrix}
0 & 0 \\
0 & I
\end{pmatrix}
:= E_2 \quad.
$$

Hence $\mathrm{Ad}u \circ \phi = \phi^1 \oplus \phi^2$ with

$$
\phi^1 := \mathrm{Ad}u \circ \phi|_{A_1} :\; A_1 \longrightarrow E_1 B E_1 \cong M_{m_1}(C(Y)) \;,
$$

$$
\phi^2 := \mathrm{Ad}u \circ \phi|_{A_2} :\; A_2 \longrightarrow E_2 B E_2 \cong M_{m_2}(C(Y)) \;,
$$

both being from a single block to a single block. The set $\text{Sp}\phi_y$ is δ-dense in X implies that the sets $\text{Sp}\phi_y^1$ and $\text{Sp}\phi_y^2$ are δ-dense in X_1 and X_2 respectively. Therefore, we need only to consider the homomorphism $\phi: A \to B$ with both A and B being single blocks.

Let $A = M_n(C(X))$, $B = M_{nq}(C(Y))$, and let $\phi: A \to B$ be a unital homomorphism. Under certain identifications of $M_n(C(X))$ with $C(X) \otimes M_n$ and of $M_{nq}(C(Y))$ with $M_q(C(Y)) \otimes M_n$, there is a homomorphism $\alpha: C(X) \to M_q(C(Y))$ such that

$$\phi \cong \alpha \otimes \text{id}_n,$$

where id_n is the identity map from M_n to itself. By definition, $\text{Sp}\phi_y = \text{Sp}\alpha_y$ for all $y \in Y$. Let us write $F \subset M_n(C(X))$ as $F = \{F^1, F^2, \cdots, F^K\}$, and write $F^k = \sum_{i,j} f_{ij}^k e_{ij}$, where e_{ij} are matrix units in $M_n(C(X))$. Set

$$G = \{f_{ij}^k; \quad i,j = 1,2,\cdots,n; k = 1,2,\cdots,K\}.$$

By Theorem 4.3, for $\varepsilon > 0$, finite set $G \subset C(X)$, and integer $N > 0$, there exists $\delta > 0$ such that, if $\text{Sp}\alpha_y(= \text{Sp}\phi_y)$ is δ-dense in X for all $y \in Y$, then there is a unital homomorphism $\beta: A \to B$ such that the following statements are true.

(i) $\|\alpha(g) - \beta(g)\| < \varepsilon$ for all $g \in G$;

(ii) There is a unitary $u \in M_q(C(Y))$ such that

$$\beta(f)(y) = u(y) \begin{pmatrix} f(\lambda_1(y))I_{l_1} & & & \\ & f(\lambda_2(y))I_{l_2} & & \\ & & \ddots & \\ & & & f(\lambda_p(y))I_{l_p} \end{pmatrix} u(y)^*$$

for all $f \in C(X)$, where $\lambda_i \in C(Y, X)$ and $l_i \geq N$, $i = 1, \cdots, p$.

Define $\psi = \beta \otimes \mathrm{id}_n : A \longrightarrow B$. Then, for $k = 1, 2, \cdots, K$,

$$\|\psi(F^k) - \alpha \otimes \mathrm{id}_n(F^k)\|$$

$$= \|(\beta \otimes \mathrm{id}_n)(\sum_{i,j} f_{ij}^k e_{ij}) - (\alpha \otimes \mathrm{id}_n)(\sum_{i,j} f_{ij}^k e_{ij})\|$$

$$= \|\sum_{i,j} (\beta(f_{ij}^k) - \alpha(f_{ij}^k))e_{ij}\| < \varepsilon n^2 .$$

In addition, for all $f = (f_{ij})_{n \times n} \in M_n(C(X))$,

$$\psi(f)(y) = \sum_{i,j} e_{ij} \otimes u(y) \begin{pmatrix} f_{ij}(\lambda_1(y))I_{l_1} & & & \\ & f_{ij}(\lambda_2(y))I_{l_2} & & \\ & & \ddots & \\ & & & f_{ij}(\lambda_p(y))I_{l_p} \end{pmatrix} u(y)^*$$

$$= (I_n \otimes u(y)) \begin{pmatrix} f(\lambda_1(y)) \otimes I_{l_1} & & & \\ & f(\lambda_2(y)) \otimes I_{l_2} & & \\ & & \ddots & \\ & & & f(\lambda_p(y)) \otimes I_{l_p} \end{pmatrix} (I_n \otimes u(y)^*).$$

□

Remark 4.11. Let $A = \lim\limits_{n\to\infty}(A_n, \phi_{n,m})$ be an injective unital inductive limit system. It follows from Proposition 2.1 of [DNNP] (or from [El1]) that, for each A_n, there is an A_m $(m > n)$ such that $\mathrm{Sp}(\phi_{n,m}^{i,j})_y$ is δ-dense in $\mathrm{Sp}A_n^i$ for each $y \in \mathrm{Sp}A_m^j$ (see Lemma 6.4 for detail). If one further supposes that $A \neq M_n(\mathbb{C})$, then for any A_n and $N > 0$, there is an M such that

$$\frac{\mathrm{rank}(\phi_{n,m}^{i,j}(\mathbf{1}_{A_n^i}))}{\mathrm{rank}(\mathbf{1}_{A_n^i})} \geq N$$

for each $m > M$.

Theorem 4.12.   Let $A = \lim\limits_{n\to\infty}(A_n, \phi_{n,m})$ be a unital inductive limit with each A_n being a finite direct sum of matrix algebras over trees and the limit algebra A being simple. Suppose that $A \neq M_k(\mathbb{C})$. Then A is approximately divisible.

PROOF: By Theorem 2.2.1, we may assume that all the homomorphisms $\phi_{n,m}$ in the inductive limit system are injective. Given finite subset $F \subset A$, $\varepsilon > 0$ and integer $N > 0$, there is an integer $n > 0$ and a finite set $F_n \subset A_n$ such that F_n and F can be paired within $\frac{\varepsilon}{2}$. (We consider $F_n \subset A_n \subset A$ because of the injectivity.) To be more clear, we denote by $\phi_{m,\infty}(f)$ the image of f in the limit algebra A for each $f \in A_m$. Then the sets F and $\phi_{n,\infty}(F_n)$ can be paired within $\frac{\varepsilon}{2}$. For $F_n \subset A_n$, $\frac{\varepsilon}{2}$ and $N > 0$, there exists $\delta > 0$ as in Theorem 4.10. By the assumption and Remark 4.11, there is an $M > n$ such that, for all $m \geq M$ and for all partial maps $\phi_{n,m}^{i,j} : A_n^i \to A_m^j$, $\mathrm{Sp}(\phi_{n,m}^{i,j})_y$ are δ-dense in $\mathrm{Sp}A_n^i$ for all $y \in \mathrm{Sp}A_m^j$, and

$$\frac{\mathrm{rank}(\phi_{n,m}^{i,j}(\mathbf{1}_{A_n^i}))}{\mathrm{rank}(\mathbf{1}_{A_n^i})} \geq N .$$

Applying Theorem 4.10 to the homomorphism $\phi_{n,m} : A_n \to A_m$, we get a unital homomorphism $\psi : A_n \to A_m$ such that each of its partial map

$$\psi^{i,j} : A_n^i \longrightarrow \phi_{n,m}^{i,j}(\mathbf{1}_{A_n^i})A_m^j\phi_{n,m}^{i,j}(\mathbf{1}_{A_n^i})$$

is of the form

$$\psi^{i,j}(f)(y) = u(y)\begin{pmatrix} f(\lambda_1(y))I_{l_1} & & & \\ & f(\lambda_2(y))I_{l_2} & & \\ & & \ddots & \\ & & & f(\lambda_p(y))I_{l_p} \end{pmatrix}u(y)^*$$

for all $f \in A_n^i$ and $y \in \mathrm{Sp}A_m^j$, with each $l_i \geq N$ and

$$\|\psi^{i,j}(f) - \phi_{n,m}^{i,j}(f)\| < \frac{\varepsilon}{2}$$

for all $f \in G_n^i$, where $G_n^i = \pi_i(F_n) \subset A_n^i$ and $\pi_i : A_n^1 \oplus \cdots \oplus A_n^{k_n} \to A_n^i$ is the projection map, $i = 1, 2, \cdots, k_n$. Define

$$(A_m^{i,j})_0 = M_{l_1}(\mathbb{C}) \oplus M_{l_2}(\mathbb{C}) \cdots \oplus M_{l_p}(\mathbb{C}).$$

Notice that

$$\text{rank}\phi_{n,m}^{i,j}(\mathbf{1}_{A_n^i}) = \text{rank}(\mathbf{1}_{A_n^i}) \cdot \sum_{t=1}^{p} l_t \ .$$

Then

$$(A_m^j)_0 := \oplus_{i=1}^{k_n}(A_m^{i,j})_0$$

is a unital finite dimensional sub-algebra of A_m^j and hence

$$(A_m)_0 := \oplus_{j=1}^{k_m}(A_m^j)_0$$

is a unital finite dimensional sub-algebra of A_m. Define

$$A_0 = \phi_{m,\infty}((A_m)_0) \qquad \text{and} \qquad F_0 = \phi_{m,\infty}(\psi(F_n)) \ .$$

Then A_0 is a finite dimensional unital sub-algebra of A, with the size of each of its direct summands being at least N. Furthermore, F_0 commutes with A_0 since $\psi(F_n)$ commutes with $(A_m)_0$.

For any $f \in F$, there is a $g \in F_n$ such that

$$\text{dist}(f, \phi_{m,\infty}(\phi_{n,m}(g))) < \frac{\varepsilon}{2}.$$

On the other hand,

$$\|\psi(g) - \phi_{n,m}(g)\| < \frac{\varepsilon}{2}$$

for all $g \in F_n$. Hence

$$\|\phi_{m,\infty}(\psi(g)) - \phi_{m,\infty}(\phi_{n,m}(g))\| < \frac{\varepsilon}{2}$$

for all $g \in F_n$. And hence

$$\text{dist}(f, \phi_{m,\infty}(\psi(g))) < \varepsilon.$$

for all $g \in F_n$. Notice that $\phi_{m,\infty}(\psi(g)) \in F_0$. We conclude that

$$\text{dist}(f, F_0) < \varepsilon$$

for all $f \in F$.

$\square$

Chapter 5

Uniqueness Theorem

In this chapter, we will prove one of the main technical theorems —Uniqueness Theorem, which will play an important role in the intertwining argument.

5.1. First, we define the "test functions" introduced in [Su]. Let $M > 1$ be a positive number, X be a compact metric space, and $\omega \subset X$ be a closed subset. Define the test function $\chi_{\omega,M} \in C(X)$ (associated to the closed set ω and the positive number M) as the following:

$$\chi_{\omega,M} = \begin{cases} 1 & \text{if } x \in \omega \\ 1 - M\operatorname{dist}(x,\omega) & \text{if } \operatorname{dist}(x,\omega) \leq \frac{1}{M} \\ 0 & \text{if } \operatorname{dist}(x,\omega) \geq \frac{1}{M}. \end{cases}$$

Notice that $\operatorname{supp}(\chi_{\omega,M}) = \overline{B_{\frac{1}{M}}(\omega)}$.

5.2. Suppose that X is a graph and each edge of X has length 1. Let r be a positive integer. Divide each edge of X into r sub-intervals of equal length. Let $\mathcal{I}_r = \{I_1, I_2, \cdots, I_K\}$ be the collection of all closed sub-intervals from the above division. It is obvious that $\mathcal{I}_r$ is a finite set.

Consider any subset $\{I_{i_1}, I_{i_2}, \cdots, I_{i_k}\} \subset \mathcal{I}_r$. Set

$$\omega = I_{i_1} \bigcup I_{i_2} \bigcup \cdots \bigcup I_{i_k} \ .$$

Then ω is a closed subset of X. Let $\mathcal{L}_r(X) \subset C(X)$ denote the collection of all test functions of form $\chi_{\omega,r}$, where ω is of the above form. Apparently $\mathcal{L}_r(X)$ is a finite set.

5.3. Let $A = \oplus_{i=1}^k M_{m_i}(C(X_i))$. Let $\mathcal{L}_r(A)$ denote $\oplus_{i=1}^k \mathcal{L}_r(X_i)$, where $\mathcal{L}_r(X_i)$ is defined as in 5.2. Note that

$$\mathcal{L}_r(A) = \oplus_{i=1}^k \mathcal{L}_r(X_i) \subseteq \oplus_{i=1}^k C(X_i) = AffTA.$$

5.4. Let $\phi : C(X) \to M_q(C(Y))$ be a unital homomorphism. Suppose that $f \in C(X)$ satisfies the following conditions:

$$0 \leq f \leq 1; \qquad f|_\omega = 1; \qquad f|_{X \backslash \Omega} = 0 \ ,$$

where $\omega \subset X$ is closed and Ω is open with $\omega \subset \Omega \subset X$. It is easily seen that for any $y \in Y$,

$$|\mathrm{Sp}\phi_y \bigcap \omega| \leq q \cdot (AffT\phi(f)(y)) \leq |\mathrm{Sp}\phi_y \bigcap \Omega|$$

(see 1.9).

Recall that, for any subset E of $Sp\phi_y$ or $Sp\psi_y$, we use $|E|$ to denote the number of the elements of E, counting the multiplicities.

Lemma 5.5. *Let X be a graph and Y be an arbitrary compact metric space. Suppose that homomorphisms $\phi, \psi : C(X) \to M_q(C(Y))$, $\delta > 0$, and positive integer r satisfy the following conditions:*

(1) For any open interval T with length $\dfrac{1}{r}$, and for each $y \in Y$,

$$|Sp\phi_y \bigcap T| \geq \delta|Sp\phi_y| = \delta q \ ,$$

$$|Sp\psi_y \bigcap T| \geq \delta|Sp\psi_y| = \delta q \ ;$$

(2) $\|AffT\phi(f) - AffT\psi(f)\| < \delta$ for all $f \in \mathcal{L}_r(X)$.

It follows that for each $y \in Y$, $Sp\phi_y$ and $Sp\psi_y$ can be paired within $\dfrac{3}{r}$.

PROOF: We will apply Marriage Lemma of [HV] (see also §2 of [Su]). For any fixed point $y \in Y$ and any subset $H \subset Sp\phi_y$, let $\omega_H \subset X$ be the closed set

$$\omega_H = \bigcup \{I_k; \ H \cap I_k \neq \varnothing, I_k \in \mathcal{I}_r\}.$$

It is obvious that $\omega_H \subseteq \overline{B_{\frac{1}{r}}(H)}$.

Note that $\mathrm{supp}\chi_{\omega_H,r} \subseteq \overline{B_{\frac{1}{r}}(\omega_H)}$. Set $\chi = \chi_{\omega_H,r} \in \mathcal{L}_r(X)$. By 5.4 and the condition (2) above,

$$|H| \leq q \cdot (AffT\phi(\chi)(y)) \leq q \cdot (AffT\psi(\chi)(y)) + q\delta.$$

Now we prove $|H| \leq |\mathrm{Sp}\psi_y \cap B_{\frac{2}{r}}(\omega_H)|$ as follows. If $\overline{B_{\frac{1}{r}}(\omega_H)} = X$, then $\mathrm{Sp}\psi_y \cap B_{\frac{2}{r}}(\omega_H) = \mathrm{Sp}\psi_y$. Hence $|H| \leq |\mathrm{Sp}\psi_y \cap B_{\frac{2}{r}}(\omega_H)| = q$.

If $\overline{B_{\frac{1}{r}}(\omega_H)} \neq X$, then there is at least one open interval $T \subset B_{\frac{2}{r}}(\omega_H)$ with length $\frac{1}{r}$ satisfying $T \cap \overline{B_{\frac{1}{r}}(\omega_H)} = \varnothing$. Therefore, by (1)

$$|\mathrm{Sp}\psi_y \cap \overline{B_{\frac{1}{r}}(\omega_H)}| + q\delta \leq |\mathrm{Sp}\psi_y \cap B_{\frac{2}{r}}(\omega_H)|.$$

Hence

$$
\begin{aligned}
|H| &\leq q \cdot (AffT\psi(\chi)(y)) + q\delta \\
&\leq |\mathrm{Sp}\psi_y \cap \overline{B_{\frac{1}{r}}(\omega_H)}| + q\delta \\
&\leq |\mathrm{Sp}\psi_y \cap B_{\frac{2}{r}}(\omega_H)| \, .
\end{aligned}
$$

Combining with $\omega_H \subset \overline{B_{\frac{1}{r}}(H)}$, we see that $B_{\frac{2}{r}}(\omega_H) \subseteq B_{\frac{3}{r}}(H)$. That is, $|H| \leq |\mathrm{Sp}\psi_y \cap B_{\frac{3}{r}}(H)|$.

Similarly, if H is any subset of $\mathrm{Sp}\psi_y$, then $|H| \leq |\mathrm{Sp}\phi_y \cap B_{\frac{3}{r}}(H)|$. The conclusion follows from Marriage Lemma of [HV].

$\square$

Theorem 5.6. *Let X be an arbitrary tree. For any $\varepsilon > 0$ and any finite subset $F \subset C(X)$, there is a positive integer r (and $\mathcal{L}_r(X) \subseteq AffT(C(X)) = C(X)$) such that, if $\delta > 0$ and two unital homomorphisms $\phi, \psi : C(X) \to M_q(C(Y))$, where Y is an arbitrary tree, satisfy the following conditions:*

(1) For each open interval $T \subset X$ of length $\frac{1}{r}$,

$$|Sp\phi_y \bigcap T| \geq \delta |Sp\phi_y| = \delta q$$

$$|Sp\psi_y \bigcap T| \geq \delta |Sp\psi_y| = \delta q ;$$

(2) $\|AffT\phi(h) - AffT\psi(h)\| < \delta$ for all $h \in \mathcal{L}_r(X)$,

then there is a unitary $u \in M_q(C(Y))$ with $\quad \|\phi(f) - u\psi(f)u^*\| < \varepsilon \quad$ for all
$f \in F \subset C(X)$.

PROOF: By Corollary 3.6, there is an $\eta > 0$ such that, if unital homomorphisms $\phi, \psi : C(X) \to M_q(C(Y))$, where Y is an arbitrary tree, satisfy the condition that $Sp\phi_y$ and $Sp\psi_y$ can be paired within η for each $y \in Y$, then for a certain unitary $u \in M_q(C(Y))$,

$$\|\phi(f) - u\psi(f)u^*\| < \varepsilon$$

for all $f \in F \subset C(X)$. The theorem follows from Lemma 5.5 by taking $\frac{3}{r} < \eta$.

$\square$

Remark 5.7. From the proofs of Lemma 5.5 and Theorem 5.6, we know that the conclusion of Theorem 5.6 holds for any $r_1 \geq r$ in place of r.

5.8. If two unital homomorphisms $\phi, \psi : M_k(C(X)) \to A$ (A is an arbitrary unital C^*-algebra) satisfy

$$\phi(e_{ij}) = \psi(e_{ij})$$

for all matrix units e_{ij}, then there is an identification of A with $M_k(pAp)$, where $p = \phi(e_{11}) = \psi(e_{11})$, such that

$$\phi = \phi_1 \otimes \mathrm{id}_k, \qquad \text{and} \qquad \psi = \psi_1 \otimes \mathrm{id}_k,$$

where $\phi_1, \psi_1 : C(X) \to pAp$ are defined by $\phi_1 = \phi|_{e_{11}M_k(C(X))e_{11}}$ and $\psi_1 = \psi|_{e_{11}M_k(C(X))e_{11}}$ respectively, and id_k is the identity map from M_k to itself.

5.9. Let $G \subset C(X)$ be a finite set of generators. For any finite set $F \subset M_k(C(X))$ and $\varepsilon > 0$, there is an $\varepsilon_1 > 0$ such that if $\phi, \psi : M_k(C(X)) \to A$ satisfy the following two conditions:

(1) $\phi(e_{ij}) = \psi(e_{ij})$ for all matrix units of $M_k(C(X))$;

(2) $\|\phi(e_{11}(g \cdot \mathbf{1}_k)e_{11}) - \psi(e_{11}(g \cdot \mathbf{1}_k)e_{11})\| < \varepsilon_1$ for all $g \in G \subset C(X)$, then

$\|\phi(f) - \psi(f)\| < \varepsilon$ for all $f \in F$.

5.10. If Y is a graph, then any projection $p \in M_l(C(Y))$ is a trivial projection. Therefore, $pM_l(C(Y))p$ is isomorphic to $M_{\mathrm{rank}(p)}(C(Y))$.

If B is a direct sum of matrix algebras over graphs, then two projections $p, q \in B$ define the same element in $K_0(B)$, if and only if they are unitarily equivalent to each other. From this fact, it is routine to prove the following (see Lemma 2.3 of [Br]): If $A = \oplus_{k=1}^a M_{n_k}(\mathbb{C})$, and if $\phi, \psi : A \to B$ are two homomorphisms with condition $K_0(\phi) = K_0(\psi)$, then there is a unitary $U \in B$ such that

$$\phi = \mathrm{Ad}U \circ \psi.$$

5.11. Suppose that

$$\phi, \psi : \oplus_{k=1}^{a} M_{n_k}(C(X_k)) \longrightarrow B ,$$

where X_k are arbitrary finite CW complexes and B is a direct sum of matrix algebras over graphs. If $K_0(\phi) = K_0(\psi)$, then there is a unitary $U \in B$, such that

$$\phi(e_{ij}^{k}) = (\mathrm{Ad}U \circ \psi)(e_{ij}^{k})$$

or all matrix units e_{ij}^{k} of each block $M_{n_k}(C(X_k))$.

This statement follows from 5.10, by restricting ϕ and ψ on

$$\oplus_{k=1}^{a} M_{n_k}(\mathbb{C}) \subset \oplus_{k=1}^{a} M_{n_k}(C(X_k)) .$$

5.12. Recall from the introduction that the following statements are true:

(i) If $\phi_1 : C(X) \to M_l(C(Y))$ and $\phi = \phi_1 \otimes \mathrm{id}_k : M_k(C(X)) \to M_{lk}(C(Y))$, then

$$AffT\phi = AffT\phi_1 : C(X) \longrightarrow C(Y);$$

(ii) Suppose that $\phi : A \to B$ is a homomorphism. Then for any unitary $u \in B$,

$$AffT\phi = AffT(\mathrm{Ad}u \circ \phi).$$

5.13. A unital homomorphism $\phi : C(X) \to M_q(C(Y))$ is said to have property $\mathrm{sdp}(r, \delta)$ (the **spectrum distribution property** with respect to r and δ) if

$$|\mathrm{Sp}\phi_y \bigcap T| \geq \delta |\mathrm{Sp}\phi_y|$$

for any $y \in Y$ and any open interval $T \subset X$ with length $\dfrac{1}{r}$.

Let Y be a graph. Then any unital homomorphism $\phi : M_k(C(X)) \to M_l(C(Y))$ can be identified with $\phi_1 \otimes \mathbf{1}_k$, where

$$\phi_1 = \phi|_{e_{11}M_k(C(X))e_{11}} : C(X) \longrightarrow \phi(e_{11})M_l(C(Y))\phi(e_{11}) \quad (\cong M_{\frac{l}{k}}(C(Y)))$$

for a certain identification of $M_l(C(Y))$ with $[\phi(e_{11})M_l(C(Y))\phi(e_{11})] \otimes M_k$. We say that ϕ has property $\mathrm{sdp}(r, \delta)$ if ϕ_1 has property $\mathrm{sdp}(r, \delta)$.

Suppose that $A = \oplus_{k=1}^{a} M_{n_k}(C(X_k))$ and $B = \oplus_{l=1}^{b} M_{m_l}(C(X_l))$, and that $\phi : A \to B$ is a unital homomorphism. We say that ϕ has property $\mathrm{sdp}(r, \delta)$ if each of its partial maps

$$\phi^{k,l} : M_{n_k}(C(X_k)) \longrightarrow \phi^{k,l}(\mathbf{1}_{n_k})M_{m_l}(C(X_l))\phi^{k,l}(\mathbf{1}_{n_k})$$

has property $\mathrm{sdp}(r, \delta)$.

Theorem 5.14. *Let $A = \oplus_{k=1}^{n} M_{n_k}(C(X_k))$, where X_k are trees. Let $F \subset A$ be a finite subset. For any $\varepsilon > 0$, there is an $r > 0$ (and the finite set $\mathcal{L}_r(A) \subset AffTA$) such that the following statement holds.*

If a number $\delta > 0$, and two unital homomorphisms

$$\phi, \ \psi : \ A \longrightarrow B = \oplus_{l=1}^m M_{m_l}(C(Y_l)) \ ,$$

where Y_l are arbitrary trees, satisfy the following three conditions:

(i) Both ϕ and ψ have property $sdp(r, \delta)$;

(ii) $\|AffT\phi(h) - AffT\psi(h)\| < \delta$, for all $h \in \mathcal{L}_r(A)$;

(iii) $K_0\phi = K_0\psi$, then there is a unitary $U \in B$ such that

$$\|\phi(f) - U\psi(f)U^*\| < \varepsilon$$

for all $f \in F$.

PROOF: Let $F_k \in M_{n_k}(C(X_k))$ be the set consisting of those elements which are the images of elements in F under the projection map from A onto its k-th block. Then $F \subset \oplus_{k=1}^n F_k$. Let G_k be a finite set of generators of $C(X_k)$. For F_k, G_k, and $\varepsilon > 0$, we can find $\varepsilon_k > 0$, as the ε_1 in 5.9. For ε_k, $G_k \subset C(X_k)$, one can find a positive integer r_k as the r in Theorem 5.6. Set $r = \max\{r_k\}$. One can check that the conclusion of the theorem holds for r (and $\mathcal{L}_r(A)$), by 5.7–5.12. (By 5.11, and (ii) of 5.12, we can assume that ϕ and ψ are equal on all the matrix units. Then by 5.8 and 5.9, we only need to consider $\phi^{k,l}$ and $\psi^{k,l}$ on $e_{11}^k M_{n_k}(C(X_k))e_{11}^k$. Actually, we only need to compare them on $G_k \subset e_{11}^k M_{n_k}(C(X_k))e_{11}^k = C(X_k)$. By the first part of 5.10, those restrictions are exactly homomorphisms from $C(X_k)$ to an algebra of a single block like $M_t(C(Y_l))$. Notice the construction of r_k and the equality $r = \max\{r_k\}$. The conclusion follows from Theorem 5.6.)

$\square$

Chapter 6

Existence Theorem and

Classification

6.1. Suppose that A and B are unital C^*-algebras and $\xi : TB \to TA$ is an affine map between the Choquet simplices. That is, ξ is a topological map between the compact metrizable spaces which keeps the affine linear combination:

$$\xi(\sum \alpha_i \tau_i) = \sum \alpha_i \xi(\tau_i)$$

for any $\tau_1, \tau_2, \cdots, \tau_n \in TB$ and $\alpha_1, \alpha_2, \cdots, \alpha_n \in \mathbb{R}_+$ with $\sum \alpha_i = 1$. Then ξ induces a linear map

$$\xi^* : AffTA \longrightarrow AffTB$$

defined by

$$\xi^*(f)(\tau) = f(\xi(\tau))$$

for all $f \in AffTA$ and $\tau \in TB$.

It is obvious that

$$\xi^*(AffTA_+) \subset AffTB_+, \qquad \xi^*(1) = 1.$$

Hence ξ induces a positive unital linear map (or scaled ordered map) from $AffTA$ to $AffTB$. In addition, if ξ is an isomorphism, then so is ξ^*.

For any $\tau \in TA$ and any element $x \in K_0(A)$, one can define $\tau(x)$ as follows. Let $x = [p] - [q]$, where $p, q \in M_n(A)$ are projections. Define

$$\tau(x) = \sum_i \tau(p_{ii}) - \sum_i \tau(q_{ii}) \in \mathbb{R} \subset \mathbb{C}.$$

Then τ induces a group homomorphism from $K_0(A)$ to $\mathbb{R}$ by $x(\tau) := \tau(x)$. By this way, each element $x \in K_0(A)$ induces an affine linear map from TA to $\mathbb{R}$, and therefore, defines an element in $AffTA$. This gives us a map $\sigma : K_0(A) \to AffTA$.

Let $\alpha : K_0A \to K_0B$ be a scaled ordered homomorphism, and $\xi : TB \to TA$ is an affine map. We say α and ξ are **compatible** if

$$\tau(\alpha(x)) = (\xi(\tau))(x) \qquad\qquad (*)$$

for all $x \in K_0A$ and $\tau \in TB$.

It is evident that, α and ξ are compatible if and only if the following diagram commutes:

$$
\begin{array}{ccc}
K_0A & \xrightarrow{\alpha} & K_0B \\
\sigma \downarrow & & \sigma \downarrow \\
AffTA & \xrightarrow{\xi^*} & AffTB.
\end{array}
$$

In the rest of this chapter, we will only use the map from $AffTA$ to $AffTB$. So instead of ξ^*, we will use ξ to denote this map.

6.2. Suppose that $A = \lim\limits_{n \to \infty}(A_n, \phi_{n,m})$ and $B = \lim\limits_{n \to \infty}(B_n, \psi_{n,m})$ are simple C^*-algebras, where the algebras $A_n = \oplus_{i=1}^{k_n} M_{[n,i]}(C(X_{n,i}))$ and $B_n = \oplus_{j=1}^{l_n} M_{[n,j]}(C(Y_{n,j}))$, and the spaces $X_{n,i}$ and $Y_{n,j}$ are trees. By Theorem 2.2.1, without loss of generality, we assume that all $\phi_{n,m}$ and $\psi_{n,m}$ are injective.

Suppose that $\alpha : K_0 A \to K_0 B$ is a scaled ordered isomorphism with inverse α^{-1}, and that $\xi : AffTA \to AffTB$ is a scaled ordered isomorphism between ordered complex Banach spaces with inverse ξ^{-1}. We assume that α and ξ are compatible in the sense of $(*)$ in 6.1. The main purpose of this chapter is to lift the maps into finite stages, that is, define $\alpha_n : K_0 A_n \to K_0 B_m$ and $\xi_n : AffTA_n \to AffTB_m$ (pass to sub-sequences) with certain properties, and to find a homomorphism $\phi_n : A_n \to B_m$ with $K_0\phi_n = \alpha_n$ and $AffT\phi_n = \xi_n$, (here one has to pass to sub-sequences again). This is called existence theorem in Elliott's frame work of the classification theory. Finally, we will prove the classification theorem.

The following lemma is well known. (See [El] [El1].)

Lemma 6.3. *There are sub-sequences $A_{n_1}, A_{n_2}, \cdots, A_{n_i}, \cdots; B_{m_1}, B_{m_2}, \cdots, B_{m_i}, \cdots$ and scaled ordered K_0 maps $\alpha_i : K_0 A_{n_i} \to K_0 B_{m_i}$, $\beta_i : K_0 B_{m_i} \to K_0 A_{n_{i+1}}$ such* *that the following diagram commutes,*

That is, for all $i = 1, 2, \cdots$,

$$\beta_i \circ \alpha_i = K_0 \phi_{n_i, n_{i+1}} ,$$

$$\alpha_{i+1} \circ \beta_i = K_0 \psi_{m_i, m_{i+1}} ,$$

$$\alpha \circ K_0 \phi_{n_i, \infty} = K_0 \psi_{m_i, \infty} \circ \alpha_i ,$$

$$\alpha^{-1} \circ K_0 \psi_{m_i, \infty} = K_0 \phi_{n_{i+1}, \infty} \circ \beta_i .$$

For convenience, from now on, we will assume $n_i = i$ and $m_i = i$.

Lemma 6.4. *Let* $A = \lim\limits_{n \to \infty} (A_n, \phi_{n,m})$ *be as in 6.2. The following statements are true.*

(a) For any small open set $T \subset SpA_n$, *there is an integer* N *(large enough) such that, if* $m \geq N$, *then* $Sp(\phi_{n,m})_y \bigcap T \neq \varnothing$ *for all* $y \in SpA_m$.

Furthermore, for any $\eta > 0$, *there is an integer* N *such that, if* $m \geq N$, *then* $Sp(\phi_{n,m})_y$ *is an* η-*dense subset of* SpA_n *for each* $y \in SpA_m$.

(b) For any integer r *and* A_n, *there exist* $\delta > 0$ *and* $N > 0$ *such that, for all* $m \geq N$, $\phi_{n,m} : A_n \to A_m$ *has property* $sdp(r, \delta)$ *(see 5.13).*

PROOF: (a) The first statement of (a) is well known (see [DNNP] or [Ell]). The second statement of (a) can be proved as follows. Choose an integer $M > \frac{2}{\eta}$.

Consider the finite collection of all the open intervals of form $(\frac{l}{M}, \frac{l+1}{M})$ inside each edge of SpA_n. There is an N such that, if $m \geq N$, then

$$Sp(\phi_{n,m})_y \bigcap (\tfrac{l}{M}, \tfrac{l+1}{M}) \neq \varnothing$$

for all the above intervals. The statement follows from the fact that any open interval of length η contains an interval of form $(\frac{l}{M}, \frac{l+1}{M})$.

(b) Apply the second part of (a) to $\eta < \dfrac{1}{r}$. Then there is an N such that

$$Sp(\phi_{n,N})_y \bigcap T \neq \varnothing$$

for all open intervals T of length $\dfrac{1}{r}$. Therefore, for all such T,

$$|Sp(\phi_{n,N})_y \bigcap T| \geq 1.$$

Let $\delta = \min_j \frac{1}{\text{size}A_N^j}$. Then $\delta \leq \min_{i,j} \frac{1}{|Sp\phi_{n,N}^{i,j}|}$, where $|Sp\phi_{n,N}^{i,j}| := |Sp(\phi_{n,N}^{i,j})_y|$ for any $y \in SpA_N^j$. It is evident that $\phi_{n,N}$ has property $sdp(r, \delta)$.

The lemma follows from the following remark.

$\square$

Remark 6.5. (a) If a unital homomorphism $\phi : M_k(C(X)) \to M_l(C(Y))$ has property $sdp(r, \delta)$, then $\psi \circ \phi$ has property $sdp(r, \delta)$ for any unital homomorphism $\psi : M_l(C(Y)) \to M_m(C(Z))$.

(b) If both $\phi : M_k(C(X)) \to M_{l_1}(C(Y))$ and $\psi : M_k(C(X)) \to M_{l_2}(C(Y))$ have property $sdp(r, \delta)$, then

$$\text{diag}(\phi, \psi) : M_k(C(X)) \longrightarrow M_{l_1+l_2}(C(Y))$$

has property $\text{sdp}(r, \delta)$.

Lemma 6.6. *Let $\alpha_n : K_0A_n \to K_0B_n$ and $\beta_n : K_0B_n \to K_0A_{n+1}, n = 1, 2, \cdots$ be as in Lemma 6.3. And let $\xi : AffTA \to AffTB$ be the isomorphism in 6.2 (it is compatible with $\alpha : K_0A \to K_0B$). For any A_n, any given finite set $F \subseteq AffTA_n$, and any $\varepsilon > 0$, there exist $m > n$ and a map $\xi_n : AffTA_n \longrightarrow AffTB_m$ such that, for all $f \in F$,*

$$\|(AffT\psi_{m,\infty} \circ \xi_n)(f) - (\xi \circ AffT\phi_{n,\infty})(f)\| < \varepsilon .$$

In particular, ξ_n can be chosen to be compatible with $K_0\psi_{n,m} \circ \alpha_n$.

PROOF: For $A_n = \oplus_{i=1}^{k_n} M_{[n,i]}(C(X_{n,i}))$, in order to simplify the notation, we write $k_n = s, X_{n,i} = X_i$. Note that

$$K_0A_n = \underbrace{\mathbb{Z} \oplus \cdots \oplus \mathbb{Z}}_{s-copies} ,$$

$$AffTA_n = C(X_1) \oplus \cdots \oplus C(X_s) .$$

It is obvious that the canonical map $\sigma : K_0A_n \longrightarrow AffTA_n$ is injective. Regard K_0A_n as a subset of $AffTA_n$. The inclusion map

$$K_0A_n \hookrightarrow AffTA_n$$

is defined by

$$(a_1, a_2, \cdots, a_s) \longmapsto \left(\frac{a_1}{[n,1]}, \frac{a_2}{[n,2]}, \cdots, \frac{a_s}{[n,s]} \right) .$$

For $\varepsilon_1 > 0$ (to be determined later), and the finite set F, there exists $\delta_1 > 0$ ($\delta_1 <$

$\frac{1}{4}$) such that $|x_1 - x_2| < \delta_1$ implies that $|f(x_1) - f(x_2)| < \varepsilon_1$ for all $f \in F$. For each space X_k, let $U_{k1}, U_{k2}, \cdots, U_{k\Lambda_k}$ be a finite open covering of X_k, where for each $t = 1, 2, \cdots \Lambda_k$,

$$U_{kt} = \{x \in X_k;\ \mathrm{dist}(x, x_{kt}) < \delta_1\}$$

for a certain point $x_{kt} \in X_k$. Let $\{h_{kt}\}$ be the partition of unity subordinate to $\{U_{kt}\}$. Then $\sum_{t=1}^{\Lambda_k} h_{kt} = \chi_{X_k}$, where χ_{X_k} denotes the characteristic function corresponding to the set $X_k \subset X$.

Let $\pi : AffTA_n \to \mathbb{C}^\Lambda$ and $\tilde{\pi} : \mathbb{C}^\Lambda \to AffTA_n$ be the maps defined by

$$\pi(f) = \sum_{k,t} f(x_{kt})v_{kt} \qquad \text{and} \qquad \tilde{\pi}(\sum_{k,t} a_{kt}v_{kt}) = \sum_{k,t} a_{kt}h_{kt}$$

respectively, where $\Lambda = \Lambda_1 + \Lambda_2 + \cdots + \Lambda_s$, and $\{v_{11}, v_{12}, \cdots, v_{1\Lambda_1}, \cdots, v_{s1}, \cdots, v_{s\Lambda_s}\}$ is the standard basis of $\mathbb{C}^\Lambda = \mathbb{C}^{\Lambda_1} \oplus \cdots \oplus \mathbb{C}^{\Lambda_s}$. Then for each $f \in F$,

$$\|\tilde{\pi} \circ \pi(f) - f\| = \|\sum_{k,t} f(x_{kt})h_{kt}(x) - f(x)\| < \varepsilon_1.$$

Let us consider the following diagram:

where $\bar{\xi}_n$ and ξ_n will be defined later.

Let $E_k \in AffTA_n$ be defined by

$$E_k(x) = \begin{cases} 1 & \text{if } x \in X_k \\ 0 & \text{if } x \notin X_k \end{cases}.$$

(Note that E_k can be regarded as $(0, \cdots, 0, [n, k], 0, \cdots, 0) \in K_0A_n$.) Then $\pi(E_k) = v_{k1} + \cdots + v_{k\Lambda_k}$, $k = 1, 2, \cdots, s$. Replacing v_{k1} by $\pi(E_k)$ for all k, we get another basis of $\mathbb{C}^\Lambda$:

$$\mathcal{K} := \{\pi(E_1), v_{12}, \cdots, v_{1\Lambda_1}, \pi(E_2), v_{22}, \cdots, v_{2\Lambda_2}, \cdots, \pi(E_s), v_{s2}, \cdots, v_{s\Lambda_s}\}.$$

For $\varepsilon_1 > 0$ and the finite set $\pi(F) \subset \mathbb{C}^\Lambda$, there exists $\delta_2 > 0$ such that, for any two linear maps $L_i : \mathbb{C}^\Lambda \to Z$, $i = 1, 2$, the condition

$$\|L_1(x) - L_2(x)\| < \delta_2 \qquad \text{for all } x \in \mathcal{K}$$

implies that

$$\|L_1(f) - L_2(f)\| < \varepsilon_1 \qquad \text{for all } f \in \pi(F),$$

where Z is an arbitrary linear normed space. If we choose $m > n$ large enough, then for each $\tilde{\pi}(v_{kt}) = h_{kt} \in AffTA_n$, there exists $g_{kt} \in AffTB_m$ such that

$$\|\xi \circ AffT\phi_{n,\infty}(h_{kt}) - AffT\psi_{m,\infty}(g_{kt})\| < \delta_2,$$

where $1 \leq k \leq s$ and $2 \leq t \leq \Lambda_k$. Define $\bar{\xi}_n(v_{kt}) = g_{kt}$, $2 \leq t \leq \Lambda_k, 1 \leq k \leq s$.

For $\pi(E_k) \in \mathbb{C}^\Lambda$, we know that $\tilde{\pi}(\pi(E_k)) = E_k \in K_0 A_n \subset Aff T A_n$, and that $\alpha'(E_k) \in K_0 B_m \subset Aff T B_m$, where $\alpha' = K_0 \psi_{n,m} \circ \alpha_n : K_0 A_n \to K_0 B_m$. So we define $\bar{\xi}_n(\pi(E_k)) = \alpha'(E_k)$, $k = 1, 2, \cdots, s$. Then

$$\| \xi \circ Aff T \phi_{n,\infty} \circ \tilde{\pi}(v_{kt}) - Aff T \psi_{m,\infty} \circ \bar{\xi}_n(v_{kt}) \|$$

$$= \quad \| \xi \circ Aff T \phi_{n,\infty}(h_{kt}) - Aff T \psi_{m,\infty}(g_{kt}) \|$$

$$< \quad \delta_2 \ ,$$

where $2 \le t \le \Lambda_k$, $1 \le k \le s$. And

$$\| \xi \circ Aff T \phi_{n,\infty} \circ \tilde{\pi}(\pi(E_k)) - Aff T \psi_{m,\infty}(\bar{\xi}_n(\pi(E_k))) \|$$

$$= \quad \| \xi \circ Aff T \phi_{n,\infty} \circ \tilde{\pi}(\pi(E_k)) - Aff T \psi_{m,\infty}(\alpha'(E_k)) \|$$

$$= \quad \| \xi \circ Aff T \phi_{n,\infty}(E_k) - K_0 \psi_{m,\infty}(\alpha'(E_k)) \|$$

$$= \quad \| \alpha \circ K_0 \phi_{n,\infty}(E_k) - K_0 \psi_{m,\infty}(\alpha'(E_k)) \|$$

$$= \quad 0 < \delta_2.$$

Extend the definition of $\bar{\xi}_n$ from $\mathcal{K}$ to the whole $\mathbb{C}^\Lambda$ linearly. Then

$$\| \xi \circ Aff T \phi_{n,\infty} \circ \tilde{\pi}(f) - Aff T \psi_{m,\infty} \circ \bar{\xi}_n(f) \| < \varepsilon_1$$

for any $f \in \pi(F)$.

Let $\xi_n = \bar{\xi}_n \circ \pi$. Then for each $f \in F$,

$$\|(AffT\psi_{m,\infty} \circ \xi_n)(f) - (\xi \circ AffT\phi_{n,\infty})(f)\|$$

$$\leq \|(AffT\psi_{m,\infty} \circ \xi_n)(f) - (\xi \circ AffT\phi_{n,\infty} \circ \tilde{\pi} \circ \pi)(f)\|$$

$$+\|(\xi \circ AffT\phi_{n,\infty} \circ \tilde{\pi} \circ \pi)(f) - (\xi \circ AffT\phi_{n,\infty})(f)\|$$

$$\leq \|(AffT\psi_{m,\infty} \circ \bar{\xi}_n)(\pi(f)) - (\xi \circ AffT\phi_{n,\infty} \circ \tilde{\pi})(\pi(f))\|$$

$$+\|(\tilde{\pi} \circ \pi)(f) - f\|$$

$$\leq \varepsilon_1 + \varepsilon_1 = 2\varepsilon_1.$$

Notice that $K_0(A_n) \subset \text{span}\{E_1, E_2, \cdots, E_s\} \subset AffTA_n$, and that $\xi_n(E_k) = \alpha'(E_k)$ for all $k = 1, 2, \cdots, s$. The map ξ_n is compatible with the K_0 map, that is,

$$\xi_n|_{K_0 A_n} = \alpha' = K_0\psi_{n,m} \circ \alpha_n.$$

Finally, set $\varepsilon_1 = \dfrac{\varepsilon}{2}$ to end the proof.

$\square$

The following lemma is a consequence of Theorem 4.9 and Lemma 6.4.

Lemma 6.7. *Let $A = \lim\limits_{n\to\infty}(A_n, \phi_{n,m})$ be as in 6.2. For any $\varepsilon > 0$, integer $L > 0$, and any finite sets $F \subset A_n$ and $E \subseteq AffTA_n$, there is an integer N (large) such that, when $m > N$, there is a homomorphism $\tilde{\phi}_{n,m} : A_n \to A_m$ with the following properties:*

(1) $\|\tilde{\phi}_{n,m}(f) - \phi_{n,m}(f)\| < \varepsilon$ for any $f \in F$;

(2) $\|(AffT\tilde{\phi}_{n,m})(f) - (AffT\phi_{n,m})(f)\| < \varepsilon$ for any $f \in E$;

(3) For any minimal direct summand $A_m^k = M_t(C(X)) \subseteq A_m$, let $P^k : A_m \to$

A_m^k *be the projection map. There are maps*

$$\phi_l : A_m \longrightarrow M_{t_l}(C(X)) \subset M_t(C(X)) = A_m^k$$

and integers $n_l > L$, $l = 1, 2, \cdots, s$ with the properties $\sum_{l=1}^{s} t_l n_l = t$ and

$$P^k \circ \tilde{\phi}_{n,m} = diag(\underbrace{\phi_1, \cdots, \phi_1}_{n_1 - times}, \underbrace{\phi_2, \cdots, \phi_2}_{n_2 - times}, \cdots, \underbrace{\phi_s, \cdots, \phi_s}_{n_s - times})$$

for one way to identify $\oplus_l M_{t_l n_l}(C(X))$ as a unital sub-algebra of $M_t(C(X)) = A_m^k$.

PROOF: If one drops the property (2) above, then the lemma is a restatement of

Theorem 4.10 by Lemma 6.4.

Identify each element $(g_1, g_2, \cdots, g_{k_n}) \in AffTA_n = C(X_1) \oplus C(X_2) \oplus \cdots \oplus$

$C(X_{k_n})$ as

$$(g_1 \cdot \mathbf{1}_{[n,1]}, g_2 \cdot \mathbf{1}_{[n,2]}, \cdots, g_{k_n} \cdot \mathbf{1}_{[n,k_n]}) \in A_n.$$

Then $AffTA_n$ can be regarded as a subset of A_n. If one enlarges the set $F \subseteq A_n$

to contain $E \subseteq AffTA_n$ as a subset, then it is evident that (1) implies (2) in the

lemma,

□

The following result is in [El1], which was proved by applying a result in [T].

Theorem 6.8. *Let $A = \oplus_i M_{k_i}(C(X_i))$, $B = \oplus_j M_{l_j}(C(Y_j))$, where the spaces*

X_i *and Y_j are trees. Let $\alpha : K_0 A \to K_0 B$ be a homomorphism of ordered groups* preserving the class of the unit. Let $\xi : AffTA \to AffTB$ be a continuous positive linear map which is compatible with α.

It follows that for any finite set $F \subset AffTA$ and any $\varepsilon > 0$, there exists a finite family of unital homomorphisms $\{\psi_i\}_{i=1}^p$ from A to B such that

(a) Every ψ_i induces α on K_0 stage;

(b) $\|(\frac{1}{p}\sum_{i=1}^p AffT\psi_i - \xi)(f)\| < \varepsilon$ for all $f \in F$.

The following theorem is called the **Existence Theorem** which plays an important role in the proof of the main result of this paper.

Theorem 6.9. Let $A = \lim_{n\to\infty}(A_n, \phi_{n,m})$ and $B = \lim_{n\to\infty}(B_n, \psi_{n,m})$ be simple inductive limit C^*-algebras, where the algebras A_n and B_n are of the form $\oplus_s M_{n_s}(C(X_s))$ and the spaces X_s are trees. Suppose that there is a unital positive linear map $\xi : AffTA \to AffTB$ and an intertwining on K_0 stage,

$$
\begin{array}{ccccccccc}
K_0 A_1 & \longrightarrow & K_0 A_2 & \longrightarrow & K_0 A_3 & \to & \cdots & K_0 A & \\
\alpha_1 \downarrow & \overset{\beta_1}{\nearrow} & \alpha_2 \downarrow & \overset{\beta_2}{\nearrow} & \downarrow & & & \alpha^{-1} \uparrow\downarrow \alpha & \quad (*) \\
K_0 B_1 & \longrightarrow & K_0 B_2 & \longrightarrow & K_0 B_3 & \to & \cdots & K_0 B\,, &
\end{array}
$$

such that the following diagram commutes:

$$
\begin{array}{ccc}
K_0 A & \overset{\alpha}{\longrightarrow} & K_0 B \\
\sigma \downarrow & & \sigma \downarrow \\
AffTA & \overset{\xi}{\longrightarrow} & AffTB.
\end{array}
$$

It follows that for any integer $r > 0$, any finite set $E \subset AffTA_n, E \supset \mathcal{L}_r(A_n)$,

and any small number $\eta > 0$, there exist $\delta > 0, \delta < \eta$ and a map $\Lambda : A_n \to B_m$

(m large) such that:

(i) Λ *has property* $sdp(r, \delta)$;

(ii) $\|(AffT\psi_{m,\infty} \circ AffT\Lambda)(f) - (\xi \circ AffT\phi_{n,\infty})(f)\| < \dfrac{\delta}{2}$ *for all* $f \in E$;

(iii) $K_0\Lambda = K_0\psi_{n,m} \circ \alpha_n$.

PROOF: For $r > 0$, applying Lemma 6.4, there exist $\delta > 0$ and an integer $n_1 > 0$

such that ϕ_{n,n_1} has property $sdp(r, \delta)$. With this $\delta > 0$ and the finite set

$$E_1 := AffT\phi_{n,n_1}(E) ,$$

applying Lemma 6.6, there exist an integer $n_2 > n_1$ and a map $\zeta :\ AffTA_{n_1} \to$

$AffTB_{n_2}$ such that

$$\zeta|_{K_0A_{n_1}} = K_0\psi_{n_1,n_2} \circ \alpha_{n_1}$$

($K_0A_{n_1}$ is regarded as a subset of $AffTA_{n_1}$.) and

$$\|(\xi \circ AffT\phi_{n_1,\infty})(f) - (AffT\psi_{n_2,\infty} \circ \zeta)(f)\| < \frac{\delta}{8}$$

for all $f \in E_1$. Applying Theorem 6.8, there exist finitely many unital

homomorphisms $\Lambda_1, \Lambda_2, \cdots, \Lambda_\lambda : A_{n_1} \to B_{n_2}$ with the following properties:

(a) $K_0\Lambda_i = K_0\psi_{n_1,n_2} \circ \alpha_{n_1}$, $i = 1, 2, \cdots, \lambda$;

(b) $\|\frac{1}{\lambda}\sum_{i=1}^{\lambda} AffT\Lambda_i(f) - \zeta(f)\| < \dfrac{\delta}{8}$ for all $f \in E_1$.

Set $E_2 = \bigcup_{i=1}^{\lambda} AffT\Lambda_i(E_1)$. Applying Theorem 6.7, there exist $m > n_2$ and a map $\tilde{\psi} : B_{n_2} \to B_m$ with the following properties:

(c) $\|AffT\tilde{\psi}(f) - AffT\psi_{n_2,m}(f)\| < \dfrac{\delta}{8}$ for all $f \in E_2$;

(d) $K_0\tilde{\psi} = K_0\psi_{n_2,m}$;

(e) For any minimal direct summand $B_m^k = M_t(C(Y_k)) \subseteq B_m$ and the projection $P^k : B_m \to B_m^k$, there are maps

$$\Pi_s : B_{n_2} \to M_{t_s}(C(Y_k)) \subseteq B_m^k \subseteq B_m$$

and integers

$$m_s \geq L := \lambda \cdot ([\dfrac{16}{\delta}] + 1), \qquad s = 1, 2, \cdots, p$$

such that $\sum_{s=1}^p t_s m_s = t$ and the map $\tilde{\psi}^k := P^k \circ \tilde{\psi} : B_{n_2} \longrightarrow B_m^k$ is of the form

$$\tilde{\psi}^k = \mathrm{diag}(\underbrace{\Pi_1, \cdots, \Pi_1}_{m_1-copies}, \cdots, \underbrace{\Pi_p, \cdots, \Pi_p}_{m_p-copies}) .$$

We will define a single map $\tilde{\Lambda} : A_{n_1} \to B_m$ satisfying

$$K_0\tilde{\Lambda} = K_0\psi_{n_2,m} \circ K_0\Lambda_i, \qquad i = 1, 2, \cdots, \lambda$$

and

$$\|AffT\tilde{\Lambda}(f) - \dfrac{1}{\lambda}\sum_{i=1}^\lambda AffT\tilde{\psi} \circ AffT\Lambda_i(f)\| < \dfrac{\delta}{8}$$

for all $f \in AffT A_{n_1}, \|f\| \leq 1$.

We need only to construct the map $\tilde{\Lambda}^k : A_{n_1} \to B_m^k$ for each summand B_m^k of

B_m. Define

$$\tilde{\Lambda}^k = \mathrm{diag}(\underbrace{\mathcal{M}_1, \cdots, \mathcal{M}_1}_{[\frac{m_1}{\lambda}]}, \mathcal{M}'_1, \underbrace{\mathcal{M}_2, \cdots, \mathcal{M}_2}_{[\frac{m_2}{\lambda}]}, \mathcal{M}'_2, \cdots, \underbrace{\mathcal{M}_p, \cdots, \mathcal{M}_p}_{[\frac{m_p}{\lambda}]}, \mathcal{M}'_p),$$

where $\mathcal{M}_s$ and $\mathcal{M}'_s$ are defined as follows:

$$\mathcal{M}_s = \begin{pmatrix} \Pi_s \circ \Lambda_1 & & & \\ & \Pi_s \circ \Lambda_2 & & \\ & & \ddots & \\ & & & \Pi_s \circ \Lambda_\lambda \end{pmatrix},$$

$$\mathcal{M}'_s = \begin{pmatrix} \Pi_s \circ \Lambda_1 & & & \\ & \Pi_s \circ \Lambda_2 & & \\ & & \ddots & \\ & & & \Pi_s \circ \Lambda_{m_s - \lambda[\frac{m_s}{\lambda}]} \end{pmatrix}.$$

Then for $f \in AffTA_{n_1}$, $\|f\| \leq 1$,

$$\left\| AffT\tilde{\Lambda}^k(f) - \frac{1}{\lambda} \sum_{i=1}^{\lambda} AffT(\tilde{\psi}^k \circ \Lambda_i)(f) \right\| < \frac{\delta}{8}.$$

To see the above, one should notice that, for any unital homomorphisms $\phi_1, \phi_2, \cdots, \phi_K$,

$$AffT\mathrm{diag}(\phi_1, \phi_2, \cdots, \phi_K) = \sum_{i=1}^{K} \frac{\mathrm{size}\phi_i}{\sum_{j=1}^{K} \mathrm{size}\phi_j} AffT\phi_i,$$

and that $m_s \geq L := \lambda \cdot ([\frac{16}{\delta}] + 1)$, where $\text{size}\phi_i = \text{rank}\phi_i(\mathbf{1})$. Hence

$$\|AffT\tilde{\Lambda}(f) - (AffT\tilde{\psi} \circ (\frac{1}{\lambda} \sum_{s=1}^{\lambda} AffT\Lambda_s))(f)\| < \frac{\delta}{8}$$

and

$$\|AffT\tilde{\Lambda}(f) - (AffT\psi_{n_2,m} \circ (\frac{1}{\lambda} \sum_{s=1}^{\lambda} AffT\Lambda_s))(f)\| < \frac{\delta}{4}$$

for all $f \in E_1 \subset A_{n_1}$. Thus

$$\|AffT\tilde{\Lambda}(f) - (AffT\psi_{n_2,m} \circ \zeta)(f)\| < \frac{3\delta}{8} .$$

Set $\Lambda = \tilde{\Lambda} \circ \phi_{n,n_1}$. Then Λ satisfies the following properties:

(i) By Remark 6.5, Λ has property $\text{sdp}(r, \delta)$, since ϕ_{n,n_1} has property $\text{sdp}(r, \delta)$;

(ii) For any $f \in E$, set $f_1 = AffT\phi_{n,n_1}(f) \in E_1$. We have

$$\|(AffT\psi_{m,\infty} \circ AffT\Lambda)(f) - (\xi \circ AffT\phi_{n,\infty})(f)\|$$

$$\leq \|(AffT\psi_{m,\infty} \circ AffT\tilde{\Lambda})(f_1) - (\xi \circ AffT\phi_{n_1,\infty})(f_1)\|$$

$$\leq \|(AffT\psi_{m,\infty} \circ AffT\tilde{\Lambda})(f_1) - (AffT\psi_{n_2,\infty} \circ \zeta)(f_1)\| + \frac{\delta}{8}$$

$$\leq \|AffT\tilde{\Lambda}(f_1) - (AffT\psi_{n_2,m} \circ \zeta)(f_1)\| + \frac{\delta}{8}$$

$$\leq \frac{3\delta}{8} + \frac{\delta}{8} = \frac{\delta}{2} ;$$

(iii) By intertwining ($*$)

$$
\begin{aligned}
K_0\Lambda &= K_0\tilde{\Lambda} \circ K_0\phi_{n,n_1} \\
&= K_0\tilde{\psi} \circ K_0\Lambda_1 \circ K_0\phi_{n,n_1} \\
&= K_0\psi_{n_2,m} \circ K_0\psi_{n_1,n_2} \circ \alpha_{n_1} \circ K_0\phi_{n,n_1} \\
&= K_0\psi_{n,m} \circ \alpha_n \ .
\end{aligned}
$$

This completes the proof.

$\square$

Remark 6.10. From the proof of the above theorem, we know that the homomorphism $\Lambda : A_n \to B_m$ factors through A_{n_1},

$$
A_n \xrightarrow{\phi_{n,n_1}} A_{n_1} \xrightarrow{\tilde{\Lambda}} B_m,
$$

where the homomorphism ϕ_{n,n_1} has property $\mathrm{sdp}(r,\delta)$, and $m > n_1$. So we may assume the following to be true:

(iv) For any $s > m$, $\phi_{n,s}$ has property $\mathrm{sdp}(r,\delta)$.

The following theorem is the main result of this paper.

Theorem 6.11. *Suppose that both A and B are simple inductive limit algebras of sequences $\{A_n, \phi_{n,m}\}$ and $\{B_n, \psi_{n,m}\}$ respectively, where $\phi_{n,m} : A_n \to A_m$ and $\psi_{n,m} : B_n \to B_m$ are unital homomorphisms, and A_n and B_n are finite direct sums of matrices over the algebras of continuous functions over trees. Suppose that there* *are scaled ordered isomorphisms*

$$\alpha : K_0 A \longrightarrow K_0 B \qquad and \qquad \xi : AffTA \longrightarrow AffTB$$

which are compatible (in the sense of 6.1).

It follows that there is an isomorphism $\Phi : A \to B$ such that $K_0\Phi = \alpha$ and $AffT\Phi = \xi$.

(The theorem can be stated in terms of $\xi : TB \to TA$ instead of $\xi : AffTA \to AffTB$.)

PROOF: Let

$$A_1 \hookrightarrow A_2 \hookrightarrow \cdots \longrightarrow A$$

$$B_1 \hookrightarrow B_2 \hookrightarrow \cdots \longrightarrow B$$

be the sequences chosen as in Lemma 6.4 with

$$\alpha_n : K_0 A_n \longrightarrow K_0 B_n$$

$$\beta_n : K_0 B_n \longrightarrow K_0 A_{n+1}.$$

In order to prove the theorem, we need to construct an approximate intertwining (in the sense of [El]) of the sequences of the C^*-algebras.

In this process, we will pass to subsequences several times. Let $\nu_1, \nu_2, \cdots$ be positive numbers with $\sum_{i=1}^{\infty} \nu_i < +\infty$. We will choose subsequences

$$A_{k_1} \longrightarrow A_{k_2} \longrightarrow \cdots \quad A$$

$$B_{l_1} \longrightarrow B_{l_2} \longrightarrow \cdots \quad B$$

and maps $\Lambda^i : A_{k_i} \longrightarrow B_{l_i}$ and $\mathcal{M}^i : B_{l_i} \longrightarrow A_{k_{i+1}}$ satisfying certain conditions so

that the diagram

is an approximate intertwining. It follows from Theorem 2.1 of [El] that A and B

are isomorphic.

Let $F_i \subset A_i$, and $G_i \subset B_i$ be finite sets such that

$$F_1 \subset F_2 \subset \cdots \subset \overline{\bigcup_i F_i} = A,$$

$$G_1 \subset G_2 \subset \cdots \subset \overline{\bigcup_i G_i} = B.$$

Choose $k_1 = 1$. For ν_1 and $F_{k_1} \subset A_{k_1}$, there exists r_1 as in Theorem 5.14. For this

r_1 and $E_1 := \mathcal{L}_{r_1}(A_{k_1}) \subset AffTA_{k_1}$, by Theorem 6.9 and Remark 6.10, there exist

$\delta_1 > 0$ and $\Lambda^1 : A_{k_1} \to B_{l_1}$ such that:

(i) Λ^1 has property $sdp(r_1, \delta_1)$;

(ii) $\|(AffT\psi_{l_1,\infty} \circ AffT\Lambda^1)(f) - (\xi \circ AffT\phi_{k_1,\infty})(f)\| < \delta_1/2$ for all $f \in E_1$;

(iii) $K_0 \Lambda^1 = K_0 \psi_{k_1, l_1} \circ \alpha_{k_1}$;

(iv) For any $s \geq l_1$, $\phi_{k_1, s}$ has property sdp(r_1, δ_1).

Similarly, for $\nu_1 > 0$ and $\tilde{G}_{l_1} := G_{l_1} \bigcup \Lambda^1(F_{k_1})$, there exists $\tilde{r}_1$ as in Theorem 5.14. For $\tilde{r}_1$ and

$$H_1 := \mathcal{L}_{\tilde{r}_1}(B_{l_1}) \bigcup AffT\Lambda^1(E_1) \subset AffTB_{l_1},$$

by Theorem 6.9 and Remark 6.10, there exist $\tilde{\delta}_1 > 0$, $\tilde{\delta}_1 < \delta_1$ and $\bar{\mathcal{M}}^1 : B_{l_1} \to A_{\bar{k}_2}$ (choose $\eta = \delta_1$ in Theorem 6.9) such that:

(i') $\bar{\mathcal{M}}^1$ has property sdp$(\tilde{r}_1, \tilde{\delta}_1)$;

(ii') $\|(AffT\phi_{\bar{k}_2, \infty} \circ AffT\bar{\mathcal{M}}^1)(f) - (\xi^{-1} \circ AffT\psi_{l_1, \infty})(f)\| < \tilde{\delta}_1/2$ for all $f \in H_1$;

(iii') $K_0 \bar{\mathcal{M}}^1 = K_0 \phi_{l_1+1, \bar{k}_2} \circ \beta_{l_1}$;

(iv') For any $s \geq \bar{k}_2$, $\psi_{l_1, s}$ has property sdp$(\tilde{r}_1, \tilde{\delta}_1)$.

Now we have two maps $\bar{\mathcal{M}}^1 \circ \Lambda^1$, $\phi_{k_1, \bar{k}_2} : \quad A_{k_1} \longrightarrow A_{\bar{k}_2}$ satisfying the following conditions:

(i'') $\bar{\mathcal{M}}^1 \circ \Lambda^1$ and $\phi_{k_1, \bar{k}_2}$ have property sdp(r_1, δ_1) by (i) and (iv);

(iii'') $K_0(\bar{\mathcal{M}}^1 \circ \Lambda^1) = K_0 \phi_{k_1, \bar{k}_2}$ (by (iii) and (iii')).

In order to apply Theorem 5.14, we need condition (ii) in Theorem 5.14. First, since (ii') is true and $AffT\Lambda^1(E_1) \subset H_1$, so (ii') is true for $f \in AffT\Lambda^1(E_1)$. Hence we have

$$\|AffT(\phi_{\bar{k}_2, \infty} \circ \bar{\mathcal{M}}^1)(AffT\Lambda^1(f)) - \xi^{-1} \circ AffT\psi_{l_1, \infty}(AffT\Lambda^1(f))\| < \frac{\tilde{\delta}_1}{2}$$

for all $f \in E_1$. So

$$\|\xi \circ AffT(\phi_{\bar{k}_2,\infty} \circ \bar{\mathcal{M}}^1 \circ \Lambda^1))(f) - AffT\psi_{l_1,\infty}(AffT\Lambda^1(f))\| < \frac{\tilde{\delta}_1}{2}$$

for all $f \in E_1$. By (ii),

$$\|\xi \circ AffT\phi_{\bar{k}_2,\infty} \circ AffT(\bar{\mathcal{M}}^1 \circ \Lambda^1)(f) - \xi \circ AffT\phi_{k_1,\infty}(f)\| < \frac{\delta_1}{2} + \frac{\tilde{\delta}_1}{2}$$

for all $f \in E_1$. Therefore,

$$\|AffT\phi_{\bar{k}_2,\infty}[AffT(\bar{\mathcal{M}}^1 \circ \Lambda^1)(f) - AffT\phi_{k_1,\bar{k}_2}(f)]\| < \frac{\delta_1}{2} + \frac{\tilde{\delta}_1}{2}$$

for all $f \in E_1$. There exists $k_2 > \bar{k}_2$ such that

$$\|AffT\phi_{\bar{k}_2,k_2}[AffT(\bar{\mathcal{M}}^1 \circ \Lambda^1)(f) - AffT\phi_{k_1,\bar{k}_2}(f)]\| < \delta_1$$

for all $f \in E_1$. Thus,

$$\|AffT(\phi_{\bar{k}_2,k_2} \circ \bar{\mathcal{M}}^1 \circ \Lambda^1)(f) - AffT\phi_{k_1,k_2}(f)\| < \delta_1$$

for all $f \in E_1$. Choose

$$\mathcal{M}^1 = \phi_{k_2,\bar{k}_2} \circ \bar{\mathcal{M}}^1 : \ B_{l_1} \longrightarrow A_{k_2}.$$

Then it is easily seen that (i″) and (iii″) are still true when we replace $\bar{k}_2$ by k_2, and $\bar{\mathcal{M}}^1$ by $\mathcal{M}^1$. By the above estimation and Theorem 5.14 (and the choice of

r_1), there is a unitary u such that

$$\|\phi_{k_1,k_2}(f) - \mathrm{Ad}u(\mathcal{M}^1 \circ \Lambda^1)(f)\| < \nu_1$$

for all $f \in F_{k_1}$. Therefore, if we change $\mathcal{M}^1$ to $\mathrm{Ad}u \circ \mathcal{M}^1$, then

$$\|\phi_{k_1,k_2}(f) - (\mathcal{M}^1 \circ \Lambda^1)(f)\| < \nu_1$$

for all $f \in F_{k_1} \subset A_{k_1}$.

We need to construct the diagram step by step. Suppose that there exists the following diagram

Then we can construct Λ^{i+1} in a similar way. Choose

$$\tilde{F}_{k_{i+1}} = F_{k_{i+1}} \bigcup \mathcal{M}^i(\tilde{G}_{l_i}) \bigcup \phi_{k_i,k_{i+1}}(\tilde{F}_{k_i}).$$

For the set $\tilde{F}_{k_{i+1}}$ and the number ν_{i+1}, find r_{i+1} as in Theorem 5.14. Set

$$E_{i+1} = \mathcal{L}_{r_{i+1}}(A_{k_{i+1}}) \bigcup AffT\mathcal{M}^i(H_i).$$

By the same procedure above, we can construct $\Lambda^{i+1} : A_{k_{i+1}} \to B_{l_{i+1}}$. It should be noted that when we apply Theorem 6.9, we choose $\eta = \tilde{\delta}_i$, (i.e., δ_{i+1} should be smaller than $\tilde{\delta}_i$). In this way, we can obtain an approximately intertwining diagram

$$
\begin{array}{ccccccccc}
A_{k_1} & \longrightarrow & A_{k_2} & \longrightarrow & A_{k_3} & \longrightarrow & \cdots & \longrightarrow & A \\
\Lambda^1 \downarrow & \mathcal{M}^1 \nearrow & \Lambda^2 \downarrow & \mathcal{M}^2 \nearrow & \Lambda^3 \downarrow & \mathcal{M}^3 \nearrow & & & \\
B_{l_1} & \longrightarrow & B_{l_2} & \longrightarrow & B_{l_3} & \longrightarrow & \cdots & \longrightarrow & B
\end{array}
$$

with

$$
\|\Lambda^{i+1} \circ \mathcal{M}^i(f) - \psi_{l_i,l_{i+1}}(f)\| < \nu_i \qquad \text{for} \qquad f \in \tilde{G}_{l_i}
$$

and

$$
\|\mathcal{M}^i \circ \Lambda^i(f) - \phi_{k_i,k_{i+1}}(f)\| < \nu_i \qquad \text{for} \qquad f \in \tilde{F}_{k_i}.
$$

Hence A and B are isomorphic by Theorem 2.1 and Theorem 2.2 of [El].

□

6.12.   Example   In our main Theorem and Elliott's result, it is assumed that

$$
\phi_0 : K_0 A \longrightarrow K_0 B \qquad \text{and} \qquad \phi_T : T B \longrightarrow T A
$$

are compatible. If $\phi : A \to B$ is an isomorphism, then the induced map $K_0\phi : K_0 A \to K_0 B$ and $T\phi : T B \to T A$ are isomorphisms which are compatible. So the compatibility of ϕ_0 and ϕ_T is a necessary condition. But one may wonder that, if one only assumes the existence of the isomorphisms between $K_0 A$ and $K_0 B$, and between $T A$ and $T B$, and does not assume that they are compatible at the first place, can he still get an isomorphism between algebras A and B — are there some other isomorphisms between $K_0 A$ and $K_0 B$, and between $T A$ and $T B$, which are compatible? Generally the answer is "No". We are going to present an example based on a result of Villadsen.

Let $S(K_0A)$ denote the set of all states on K_0A (group homomorphisms from K_0A to $\mathbb{R}$, which map $(K_0A)_+$ to $\mathbb{R}_+$, and $\mathbf{1}_A$ to $\mathbf{1}$). There is a natural map $\gamma_A : TA \to S(K_0A)$ defined by

$$\gamma_A(\tau)(x) = \tau(x),$$

where $\tau \in TA$, $x \in K_0A$, and $\tau(x)$ is defined in 6.1.

Suppose that A and B are two C^*-algebras and $\phi_0 : K_0A \to K_0B$ is an ordered map taking K_0A_+ to K_0B_+, $\mathbf{1}_A$ to $\mathbf{1}_B$. Then $\phi_0 : K_0A \to K_0B$ induces a map $\phi_0^S : S(K_0B) \to S(K_0A)$ by

$$\phi_0^S(f)(x) = f(\phi_0(x))$$

for any $x \in K_0B$ and $f \in S(K_0A)$. Two maps $\phi_0 : K_0A \to K_0B$ and $\phi_T : TB \to TA$ are compatible if and only if the following diagram is commutative:

$$
\begin{array}{ccc}
TB & \xrightarrow{\gamma_B} & S(K_0B) \\
{\scriptstyle \phi_T}\downarrow & & \downarrow{\scriptstyle \phi_0^S} \\
TA & \xrightarrow{\gamma_A} & S(K_0A)
\end{array} \quad .
$$

Let $\partial_e(\Delta)$ denote the collection of all extreme points of a convex set Δ.

The main result in [V] is the following: Suppose that (G, u) is a simple scaled ordered unperforated countable Riesz group, Δ is a metrizable Choquet simplex, and $\lambda : \Delta \to S(G, u)$ is a continuous affine map with $\lambda(\partial_e \Delta) = \partial_e S(G, u)$. Then there is a simple C^*-algebra A arising as an inductive limit of $\bigoplus_{i=1}^{k_n} M_{[n,i]}(C[0,1])$

such that

$$((K_0 A, [\mathbf{1}_A]),\ TA,\ \gamma_A) = ((G, u),\ \Delta,\ \lambda).$$

Set

$$\mathbb{Z}(\frac{1}{2}) = \{\frac{m}{2^n};\ m, n \text{ are integers }\},$$

$$\mathbb{Z}(\frac{1}{3}) = \{\frac{m}{3^n};\ m, n \text{ are integers }\}.$$

Set $G = \mathbb{Z}(\frac{1}{2}) \oplus \mathbb{Z}(\frac{1}{3})$. We will write an element in G as the form (a, b), $a \in \mathbb{Z}(\frac{1}{2})$, $b \in \mathbb{Z}(\frac{1}{3})$. And set

$$G_+ = \{(a, b) \in G;\ a > 0, b > 0\} \cup \{0, 0\}, \qquad u = (1, 1) \in G.$$

It is well known that $(G,\ G_+,\ u)$ is a scaled ordered unperforated simple Riesz group.

Lemma. *If $\phi: G \to G$ is an isomorphism such that $\phi(G_+) \subset G_+$, $\phi(u) = u$, then $\phi = id$.*

PROOF: Suppose that $\phi(1, 0) = (a_1,\ b_1)$ and $b_1 \neq 0$. If m is large enough, then $\frac{b_1}{2^m} \notin \mathbb{Z}(\frac{1}{3})$. That is,

$$\phi(\frac{1}{2^m}, 0) = (\frac{a_1}{2^m},\ \frac{b_1}{2^m}) \notin G.$$

This contradiction proves that $b_1 = 0$. So $\phi(1, 0) = (a_1, 0)$. By the same reason we have $\phi(0, 1) = (0, b_2)$. Hence

$$\phi(1, 1) = (a_1, b_2).$$

Since $\phi(1,1) = (1,1)$ by assumption, $(a_1, b_2) = (1,1)$. That is,

$$\phi(1,0) = (1,0), \quad \phi(0,1) = (0,1).$$

Hence $\phi = \mathrm{id}$.

$\square$

As a consequence of the above lemma, we know that if $\phi : G \to G$ is a scaled ordered isomorphism, then $\phi^S : S(G) \to S(G)$ is the identity.

It is easy to check that $S(G)$ is a line segment with two extreme points $\theta_1, \theta_2 \in S(G)$ defined by

$$\theta_1(a,b) = a, \qquad \theta_2(a,b) = b$$

respectively.

Let Δ be a triangle with vertices (extreme points) (τ_1, τ_2, τ_3). Then Δ is a metrizable Choquet simplex.

Let $\lambda_1 : \Delta \to S(G)$ be the affine map defined by

$$\lambda_1(\tau_1) = \theta_1, \quad \lambda_1(\tau_2) = \theta_1, \quad \lambda_1(\tau_3) = \theta_2.$$

And let $\lambda_2 : \Delta \to S(G)$ be the affine map defined by

$$\lambda_2(\tau_1) = \theta_1, \quad \lambda_2(\tau_2) = \theta_2, \quad \lambda_2(\tau_3) = \theta_2.$$

It follows that $\lambda_i(\partial_e \Delta) = \partial_e(S(G))$. (Notice that λ_1 takes two extreme points to θ_1 and takes one extreme point to θ_2, and λ_2 vise versa.)

By Villadsen's theorem, there are simple C^*-algebras A and B which are inductive limits of matrix algebras over $C[0,1]$, such that

$$((K_0A, K_0A_+, \mathbf{1}_A),\ TA,\ \gamma_A)\ =\ ((G, G_+, u),\ \Delta,\ \lambda_1)$$
$$((K_0B, K_0B_+, \mathbf{1}_B),\ TB,\ \gamma_B)\ =\ ((G, G_+, u),\ \Delta,\ \lambda_2).$$

There are no isomorphisms between $TA = \Delta$ and $TB = \Delta$, and between $K_0A = G$ and $K_0B = G$, which are compatible. This can be proved as follows.

Suppose that $\phi_0 : G \to G$, $\phi_T : \Delta \to \Delta$ are isomorphisms. By the above lemma, $\phi_0^S = \mathrm{id}$. Hence the following diagram can not be commutative:

$$
\begin{array}{ccc}
\Delta & \xrightarrow{\ \lambda_1\ } & S(G) \\
{\scriptstyle \phi_T}\downarrow & & \downarrow{\scriptstyle \phi_0^S=\mathrm{id}} \\
\Delta & \xrightarrow{\ \lambda_2\ } & S(G)
\end{array}
$$

since $\phi_0^S \circ \lambda_1$ takes two extreme points of Δ to θ_1, and $\lambda_2 \circ \phi_T$ takes only one extreme point of Δ to θ_1. This proves that $A \not\cong B$.

Bibliography

[Be]   Berg, I.D. On approximation of normal operators by weighted shifts, *Michigan Math. J.* **21** (1974), 377-383.

[BKR]   Blackadar, B., Kumjian, A., and Rørdam, M., Approximately central matrix units and the structure of non-commutative tori. *K-theory* **6** (1992) 267-284.

[BBEK]   Blackadar, B., Bratteli, O., Elliott, G.A. and Kumjian, A. Reduction of real rank in inductive limits of C^*-algebras, *Math. Ann.* **292** (1992), 111-126.

[BDR]   Blackadar, B., Dadarlat, M., and Rørdam, M., The real rank of inductive limit C*-algebras, *Math. Scand.* **69** (1991), 211–216.

[Br]   Bratteli, O., Inductive limits of finite dimensional C*-algebras, *Trans. Amer. Math. Soc.* **171** (1972), 195–234.

[BrE]   Bratteli, O. and Elliott, G. A. Small eigenvalue variation and real rank zero, *Pacific J. of Math* to appear.

[BrEK]   Bratteli, O. Evans, D. and Kishimoto, A. Crossed products of totally disconnected spaces by $\mathbb{Z}_2 * \mathbb{Z}_2$, preprint.

[BrK]   Bratteli, O. and Kishimoto, A. Non-commutative sphere *III* : Irrational rotations *Communication of Math. Physics* to appear.

[DNNP]   Dadarlat, M., Nagy, G., Nemethi, A. and Pasnicu, C., Reduction of topological stable rank in inductive limits of C*-algebras, *Pacific J. of Math.* **153**, 2 (1992), 267–276.

[Di]   Dixmier, J. On some C^*-algebras considered by Glimm, *J. Functional analysis* **1** (1967) 182-203.

[El]   Elliott, G.A., On the classification of C*-algebras of real rank zero, to appear in *J. Reine Angew. Math.* **443** (1993).

[El1]   Elliott, G.A., A classification of certain simple C*-algebras, Quantum and Non-commutative Analysis (eds. by H. Araki) 373–385, 1993 Kluwer Acad. Publishers.

[El2]   Elliott, G.A., Classification of certain simple C*-algebras, II, preprint.

[EE]   Elliott, G.A. and Evans, D. The structure of the irrational rotation C*-algebra, *Annals of Math.* **3** (1993), 477–501.

[EG]   Elliott, G.A. and Gong, G. On classification of C*-algebras of real rank zero, II, *Annals of Math.*, to appear.

[EGL]   Elliott, G.A., Gong, G. and Li, L. In preparation.

[Gl]   Glimm, J. On a certain class of operator algebras, *Trans. Amer. Math. Soc.* **95** (1960), 318-340.

[HV]   Halmos, P.R. and Vaughan, H.E. Marriage problem, *American J of Math.* **72** (1950), 214-215.

[Li]   Li. L, On the classification of simple C^*-algebras: Inductive limits of matrix algebras over 1-dimensional spaces, preprint.

[P]   Putnam, I. On the topological stable rank of certain transformation group C^*-algebras, *Ergodic theory, Dynamical system* **10** (1990) 197-207.

Su]   Su, H., On the classification of C*-algebras of real rank zero: inductive limits of matrix algebras over non-Hausdorff graphs, *Mem. of A.M.S.*, to appear.

[T]   Thomsen, K. Inductive limits of interval algebras: The tracial state space, *Amer. J. Math.* 116(1994), 605–620. K. Thomsen *Inductive limits of interval algebras: The tracial state space*, Amer. J. Math., 116(1994), 605–620.

Current Address:

The Fields Institute for Research in Mathematical Sciences,
222 College Street,   Toronto, Ontario,   Canada M5T 3J1

Editorial Information

To be published in the *Memoirs*, a paper must be correct, new, nontrivial, and significant. Further, it must be well written and of interest to a substantial number of mathematicians. Piecemeal results, such as an inconclusive step toward an unproved major theorem or a minor variation on a known result, are in general not acceptable for publication. *Transactions* Editors shall solicit and encourage publication of worthy papers. Papers appearing in *Memoirs* are generally longer than those appearing in *Transactions* with which it shares an editorial committee.

As of January 31, 1997, the backlog for this journal was approximately 7 volumes. This estimate is the result of dividing the number of manuscripts for this journal in the Providence office that have not yet gone to the printer on the above date by the average number of monographs per volume over the previous twelve months, reduced by the number of issues published in four months (the time necessary for preparing an issue for the printer). (There are 6 volumes per year, each containing at least 4 numbers.)

A Copyright Transfer Agreement is required before a paper will be published in this journal. By submitting a paper to this journal, authors certify that the manuscript has not been submitted to nor is it under consideration for publication by another journal, conference proceedings, or similar publication.

Information for Authors and Editors

Memoirs are printed by photo-offset from camera copy fully prepared by the author. This means that the finished book will look exactly like the copy submitted.

The paper must contain a *descriptive title* and an *abstract* that summarizes the article in language suitable for workers in the general field (algebra, analysis, etc.). The *descriptive title* should be short, but informative; useless or vague phrases such as "some remarks about" or "concerning" should be avoided. The *abstract* should be at least one complete sentence, and at most 300 words. Included with the footnotes to the paper, there should be the 1991 *Mathematics Subject Classification* representing the primary and secondary subjects of the article. This may be followed by a list of *key words and phrases* describing the subject matter of the article and taken from it. A list of the numbers may be found in the annual index of *Mathematical Reviews*, published with the December issue starting in 1990, as well as from the electronic service e-MATH [**telnet e-MATH.ams.org** (or **telnet 130.44.1.100**). Login and password are **e-math**]. For journal abbreviations used in bibliographies, see the list of serials in the latest *Mathematical Reviews* annual index. When the manuscript is submitted, authors should supply the editor with electronic addresses if available. These will be printed after the postal address at the end of each article.

Electronically prepared papers. The AMS encourages submission of electronically prepared papers in $\mathcal{A}_\mathcal{M}\mathcal{S}$-TEX or $\mathcal{A}_\mathcal{M}\mathcal{S}$-LATEX. The Society has prepared author packages for each AMS publication. Author packages include instructions for preparing electronic papers, the *AMS Author Handbook*, samples, and a style file that generates the particular design specifications of that publication series for both $\mathcal{A}_\mathcal{M}\mathcal{S}$-TEX and $\mathcal{A}_\mathcal{M}\mathcal{S}$-LATEX.

Authors with FTP access may retrieve an author package from the Society's Internet node `e-MATH.ams.org` (130.44.1.100). For those without FTP access, the author package can be obtained free of charge by sending e-mail to pub@math.ams.org (Internet) or from the Publication Division, American Mathematical Society, P.O. Box 6248, Providence, RI 02940-6248. When requesting an author package, please specify $\mathcal{AMS}$-TEX or $\mathcal{AMS}$-LATEX, Macintosh or IBM (3.5) format, and the publication in which your paper will appear. Please be sure to include your complete mailing address.

Submission of electronic files. At the time of submission, the source file(s) should be sent to the Providence office (this includes any TEX source file, any graphics files, and the DVI or PostScript file).

Before sending the source file, be sure you have proofread your paper carefully. The files you send must be the EXACT files used to generate the proof copy that was accepted for publication. For all publications, authors are required to send a printed copy of their paper, which exactly matches the copy approved for publication, along with any graphics that will appear in the paper.

TEX files may be submitted by email, FTP, or on diskette. The DVI file(s) and PostScript files should be submitted only by FTP or on diskette unless they are encoded properly to submit through e-mail. (DVI files are binary and PostScript files tend to be very large.)

Files sent by electronic mail should be addressed to the Internet address pub-submit@math.ams.org. The subject line of the message should include the publication code to identify it as a Memoir. TEX source files, DVI files, and PostScript files can be transferred over the Internet by FTP to the Internet node e-math.ams.org (130.44.1.100).

Electronic graphics. Figures may be submitted to the AMS in an electronic format. The AMS recommends that graphics created electronically be saved in Encapsulated PostScript (EPS) format. This includes graphics originated via a graphics application as well as scanned photographs or other computer-generated images.

If the graphics package used does not support EPS output, the graphics file should be saved in one of the standard graphics formats—such as TIFF, PICT, GIF, etc.—rather than in an application-dependent format. Graphics files submitted in an application-dependent format are not likely to be used. No matter what method was used to produce the graphic, it is necessary to provide a paper copy to the AMS.

Authors using graphics packages for the creation of electronic art should also avoid the use of any lines thinner than 0.5 points in width. Many graphics packages allow the user to specify a "hairline" for a very thin line. Hairlines often look acceptable when proofed on a typical laser printer. However, when produced on a high-resolution laser imagesetter, hairlines become nearly invisible and will be lost entirely in the final printing process.

Screens should be set to values between 15% and 85%. Screens which fall outside of this range are too light or too dark to print correctly.

Any inquiries concerning a paper that has been accepted for publication should be sent directly to the Editorial Department, American Mathematical Society, P. O. Box 6248, Providence, RI 02940-6248.

Editors

This journal is designed particularly for long research papers (and groups of cognate papers) in pure and applied mathematics. Papers intended for publication in the *Memoirs* should be addressed to one of the following editors:

Ordinary differential equations, partial differential equations, and applied mathematics to JOHN MALLET-PARET, Division of Applied Mathematics, Brown University, Providence, RI 02912-9000; electronic mail: `jmp@cfm.brown.edu`.

Harmonic analysis, representation theory, and Lie theory to ROBERT J. STANTON, Department of Mathematics, The Ohio State University, 231 West 18th Avenue, Columbus, OH 43210-1174; electronic mail: `stanton@math.ohio-state.edu`.

Ergodic theory, dynamical systems, and abstract analysis to DANIEL J. RUDOLPH, Department of Mathematics, University of Maryland, College Park, MD 20742; e-mail: `djr@math.umd.edu`.

Real and harmonic analysis and geometric partial differential equations to WILLIAM BECKNER, Department of Mathematics, University of Texas at Austin, Austin, TX 78712; e-mail: `beckner@math.utexas.edu`.

Algebra and algebraic geometry to EFIM ZELMANOV, Department of Mathematics, Yale University, 10 Hillhouse Avenue, New Haven, CT 06520-8283; e-mail: `zelmanov@math.yale.edu`

Algebraic topology and cohomology of groups to STEWART PRIDDY, Department of Mathematics, Northwestern University, 2033 Sheridan Road, Evanston, IL 60208-2730; e-mail: `s_priddy@math.nwu.edu`.

Global analysis and differential geometry to CHUU-LIAN TERNG, Department of Mathematics, Northeastern University, Huntington Avenue, Boston, MA 02115-5096; e-mail: `terng@neu.edu`.

Probability and statistics to RODRIGO BAÑUELOS, Department of Mathematics, Purdue University, West Lafayette, IN 47907-1968; e-mail: `banuelos@math.purdue.edu`.

Combinatorics and Lie theory to PHILIP J. HANLON, Department of Mathematics, University of Michigan, Ann Arbor, MI 48109-1003; e-mail: `hanlon@math.lsa.umich.edu`.

Logic and universal algebra to THEODORE SLAMAN, Department of Mathematics, University of California at Berkeley, Berkeley, CA 94720; e-mail: `slaman@math.berkeley.edu`.

Number theory and arithmetic algebraic geometry to ALICE SILVERBERG, Department of Mathematics, The Ohio State University, 231 W. 18th Avenue, Columbus, OH 43210-1174; e-mail: `silver@math.ohio-state.edu`.

Complex analysis and complex geometry to DANIEL M. BURNS, Department of Mathematics, University of Michigan, Ann Arbor, MI 48109-1003; e-mail: `dburns@umich.edu`.

Algebraic geometry and commutative algebra to LAWRENCE EIN, Department of Mathematics, University of Illinois, 851 S. Morgan (M/C 249), Chicago, IL 60607-7045; email: `lawrence.man.ein@uic.edu`.

Geometric topology to JOHN LUECKE, Department of Mathematics, University of Texas at Austin, Austin, TX 78712; e-mail: `luecke@math.utexas.edu`.

All other communications to the editors should be addressed to the Managing Editor, PETER SHALEN, Department of Mathematics, University of Illinois, 851 S. Morgan (M/C 249), Chicago, IL 60607-7045; e-mail: `shalen@math.uic.edu`.

Selected Titles in This Series

(*Continued from the front of this publication*)

(See the AMS catalog for earlier titles)